Solid-Phase Organic Syntheses

SOLID-PHASE ORGANIC SYNTHESES

Volume 1

Edited by

ANTHONY W. CZARNIK

A Wiley-Interscience® Publication

JOHN WILEY & SONS, INC.

New York   Chichester   Weinheim   Brisbane   Singapore   Toronto

Library of Congress Cataloging in Publication is available.

Czarnik, Anthony W.
  Solid-Phase Organic Syntheses, Volume 1

  ISBN 0-471-31484-6

Printed in the United States of America.

10 9 8 7 6 5 4 3 2 1

CONTENTS

Preface vii

Chapter One
    2-AMINOTHIAZOLES 1

Chapter Two
    SOLID-PHASE MANNICH REACTIONS OF A
    RESIN-IMMOBILIZED SECONDARY AMINE 9

Chapter Three
    SOLID-PHASE SYNTHESIS OF UREAS ON
    MICROTUBES 15

Chapter Four
    SYNTHESIS OF p-BENZYLOXYBENZYL
    CHLORIDE RESIN 41

Chapter Five
    SOLID-PHASE MANNICH REACTIONS OF A
    RESIN-IMMOBILIZED ALKYNE 45

Chapter Six
    SOLID-PHASE SYNTHESIS OF DI-β-PEPTOIDS
    FROM ACRYLATE RESIN: N-ACETYL-N-
    BENZYL-β-ALANINYL-N-BENZYL-β-ALANINE 55

Chapter Seven
    SOLID-PHASE SYNTHESIS OF BENZOXAZOLES
    VIA MITSUNOBU REACTION 73

Chapter Eight
**N-FMOC-AMINOOXY-2-CHLOROTRITYL
POLYSTYRENE RESIN FOR HIGH THROUGH-
PUT SYNTHESIS OF HYDROXAMIC ACIDS**   **85**

Chapter Nine
**FACILE PREPARATION OF CHLORO-
METHYLARYL SOLID SUPPORTS**   **101**

Chapter Ten
PREPARATION OF AMEBA RESIN   **105**

Chapter Eleven
**AN EFFICIENT SOLID-PHASE SYNTHETIC
ROUTE TO 1,3-DISTRIBUTED 2,4 (1H, 3H)-
QUINAZOLINEDIONES SUITABLE FOR
COMBINATORIAL SYNTHESIS**   **113**

Chapter Twelve
**BACKBONE AMIDE LINKER (BAL) STRATEGY
FOR SOLID-PHASE SYNTHESIS**   **121**

Chapter Thirteen
**THE ALLYLSILYL LINKER: SYNTHESIS OF
CATALYTIC BINDING OF ALKENES AND
ALKYNES TO AND CLEAVAGE FROM
ALLYLDIMETHYLSILYL POLYSTYRENE**   **139**

Chapter Fourteen
**RESIN-BOUND ISOTHIOCYANATES AS
INTERMEDIATES FOR THE SOLID-PHASE
SYNTHESIS OF SUBSTITUTED THIOPHENES**   **149**

Author Index   **159**
Subject Index   **161**

PREFACE

All organic chemists have a working knowledge of the book series *Organic Synthesis (OS)*. This project began at my alma mater, the University of Illinois, under the directionship of Roger Adams. Adams realized that industry needed quantities of organic chemicals for its work, but there was no Aldrich yet. Thus he organized undergraduates and graduate students who worked summers to make compounds that were sold. Those reaction procedures served as the nucleus for *OS*, which evolved into a "tested" set of laboratory methods that *just plain worked*. Practicing chemists rely on *OS* to get them quickly to the point at which they can test their new idea, rather than spending weeks just getting to that point.

The newly important field of solid-phase organic synthesis desperately needs just this type of reference, in large part because much of the work occurs in industry and does not get published. If there were more cookbook-type synthetic procedures available to working synthetic chemists, this method would permeate the discovery area even faster than it currently is. *Solid-Phase Organic Syntheses (SPOS)* was created to address exactly this need.

Unlike *OS*, solid-phase methods will virtually always be invented for application in combinatorial organic synthesis. To meet these specific needs, *SPOS* procedures will focus not on multistep reactions leading to a desired final compound but rather on a single type of synthetic transformation accomplished on solid support. Because combinational syntheses will always benefit when a broad range of reactions are possible using a given method, *SPOS* procedures will have already been optimized to work with a structurally wide variety of reagents. In addition, the submittors will describe how this method works on the solid supports in common use at the time of the procedure's submission. In this way, application to small molecule library should be a rapid process.

This is the first volume in the SPOS series. Potential authors are encouraged to obtain information for making future submissions by writing to the SPOS office at SPOS@SAN.RR.COM.

Anthony W. Czarnik
San Diego, Calfornia

Solid-Phase Organic Syntheses

CHAPTER ONE

2-AMINOTHIAZOLES

Submitted by PATRICK C. KEARNEY, MONICA FERNANDEZ, MENGMENG FU, and JOHN A. FLYGARE

Tularik Inc., 2 Corporate Drive, South San Francisco, CA, USA 94080

Checked by STEPHEN SHUTTLEWORTH, AMAL WAHHAB, RICHARD WILSON, and JEANCARLO DE LUCA

BioChem Pharma, 275 Amand-Frappier Boulevard Laval, Quebee, Canada H7V

LIBRARY SYNTHESIS ROUTE

i. Fmoc-NCS, CH_2Cl_2
ii. 20% piperidine in DMF

, dioxane

95% TFA, 5% H_2O

a. Rink amide MBHA resin
b. Glycyl-rink amide MBHA resin
c.

Reductively aminated
argoGel MB-CHO resin

BUILDING BLOCKS

Amines

1. NH_3

2. 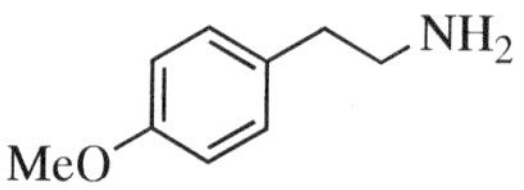   (4-methoxyphenethylamine structure, MeO–C$_6$H$_4$–CH$_2$CH$_2$–NH$_2$)

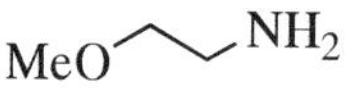

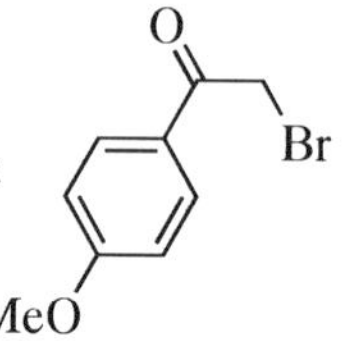

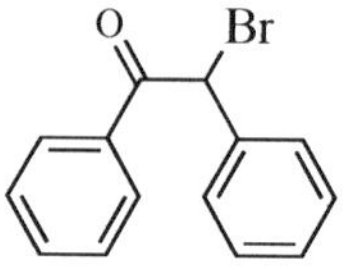

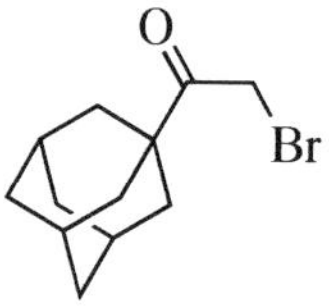

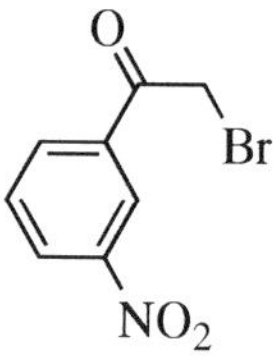

3.   MeO–CH$_2$CH$_2$–NH$_2$

4.   (cyclopentylamine structure, –NH$_2$)

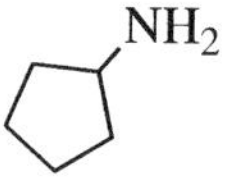

5. 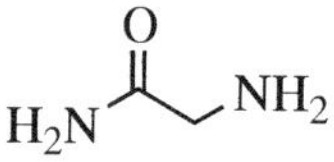 H_2N–C(=O)–CH$_2$–NH$_2$

α-Bromoketones

a   (4-methoxyphenacyl bromide structure, MeO–C$_6$H$_4$–C(=O)–CH$_2$Br)

b   (desyl bromide / 2-bromo-1,2-diphenylethanone structure)

c   (1-adamantyl bromomethyl ketone structure)

d   (3-nitrophenacyl bromide structure, NO_2)

e   (4-chlorophenacyl bromide structure, Cl)

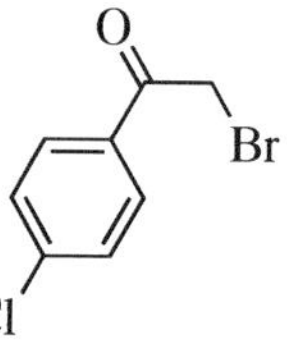

PROCEDURE

General Procedure for the Synthesis of Unsubstituted 2-Aminothiazoles (1a–e)

Rink amide MBHA resin (364 mg, 0.54 mmol/g substitution) was placed into a polypropylene reaction vessel (note 1). The resin was swollen through the addition of DMF (5 mL, 5 min, 3 ×) (note 2). The resin was then treated with a solution of 20% piperidine in DMF (5 mL, 2.5 min, 3 ×). After washing with DMF (5 mL, 30 s, 3 ×) and methylene chloride (5 mL, 30 s, 5 ×), a solution of fluorenylmethoxycarbonyl isothiocyanate (Fmoc-NCS; Note 3) in methylene chloride was applied to the resin (0.2 M, 5 mL,

20 min, 1 ×). The resin was washed with methylene chloride (5 mL, 30 s, 3 ×) and DMF (5 mL, 30 s, 3 ×) and subsequently reacted with 20% piperidine in DMF (5 mL, 2.5 min, 3 ×) to produce the resin-bound thiourea. The resin was then washed with DMF (5 mL, 30 s, 3 ×) and dioxane (5 mL, 30 s, 3 ×). The desired α-bromoketone (0.2 M) in dioxane was added (5 mL, 1 h), and the resin was washed with dioxane (5 mL, 30 s, 3 ×). The α-bromoketone addition and subsequent wash were repeated two more times. The resin was then washed with methylene chloride (5 mL, 30 s, 5 ×) and dried briefly (10 min) under a stream of nitrogen. The reaction products were cleaved with aqueous trifluoroacetic acid (TFA; 95%, 5 mL, 2 h). This eluate and two subsequent aqueous TFA washes (2.5 mL, 1 min) were collected and combined, and the solvent was removed with a Speedvac (note 4).

General Procedure for the Synthesis of *N*-Substituted Thiazoles (2a–e; 3a–e; 4a–e)

ArgoGel-MB-CHO resin (366 mg, 0.42 mmol/g substitution) was placed into an Ace pressure tube (note 5). Trimethyl orthoformate (TMOF; 5 mL) was added to the flask along with the primary amine (10 equiv.). The tube was capped and heated for 2 h at 70°C in a rotating oven (note 6), and cooled. The TMOF solution was removed with the use of a filtration cannula, and the entire process was repeated. The resin was washed with TMOF (5 mL, 1 ×) and anhydrous methanol (5 mL, 3 ×) Anhydrous methanol (5 mL) was added to the resin, followed by the addition of sodium borohydride (133 mg, 20 equiv.). After vigorous gas evolution had ceased, the tube was capped and agitated for 8 h at room temperature. The resin was then transferred to a polypropylene reaction vessel and washed with methanol (5 mL, 3 ×), methanol:water (1:1, 5 mL, 3 ×), DMF:water (1:1, 5 mL, 3 ×), DMF (5 mL, 3 ×), and methylene chloride (5 mL, 3 ×).

A modified version of this program for 2-aminothiazole synthesis was executed. In that version, the initial exposure to 20% piperidine was eliminated, and all delivered volumes were reduced to 3.75 mL. After completion of the synthesis, the resin was dried under vacuum. Aqueous TFA (95%, 5 mL) was added and the tube was heated at 50°C for 4 h (note 7). The cleavage solution and two subsequent rinses of the resin (one of 5 mL of 95% aqueous TFA and one of 5 mL of MeOH) were combined and evaporated to dryness with a Speedvac.

General Procedure for the Synthesis of *N*-Substituted Thiazoles (5a–e)

Rink amide MBHA resin (364 mg, 0.54 mmol/g substitution) was weighed out into a polyethylene reaction vessel. The resin was swollen with DMF (5 mL, 5 min, 3 ×) and subsequently treated with 20% piperidine in DMF (5 mL, 2.5 min, 3 ×). After washing with DMF (5 mL, 30 s, 5 ×), the resin was treated for 2 h with Fmoc-glycine-OH solution in DMF (0.4 M, 2.5 mL) and diisopropylcarbodiimide in DMF (0.4 M, 2.5 mL). The resin was then washed with DMF (5 mL, 30 s, 3 ×). The coupling reaction and the subsequent wash were repeated two more times. A negative ninhydrin test at this point indicated completion of the coupling reaction (note 8). The 2-aminothiazole was then constructed with the use of the corresponding bromoketone and the general procedure described above.

Description of Solid-Phase Supports

ArgoGel MB-CHO resin was purchased from Argonaut Technologies, substitution = 0.42 mmol/g, lot #104–20.

Rink amide MBHA resin was purchased from Novabiochem, substitution = 0.54 mmol/g, lot #A20678.

NOTES

1. The synthesis can be carried out manually or automated using a Symphony/Multiplex multiple peptide synthesizer or an Argonaut Nautilus.

2. Dimethylformamide (DMF), dioxane, piperidine, methylene chloride, acetonitrile, trimethyl orthoformate (TMOF), sodium borohydride, diisopropylcarbodiimide, and trifluoroacetic acid (TFA) were purchased from Aldrich Chemical Company, Inc. and used without further purification. All of the diversity reagents were purchased from Aldrich except for Fmoc-glycine-OH, which was purchased from Novabiochem.

3. Fluorenylmethoxycarbonyl isothiocyanate (Fmoc-NCS) was synthesized according to a published procedure;[1] it can also be purchased from Novabiochem.

4. Purified product was isolated with the use of a Chromatotron model 8924 apparatus (Harrison Research, Palo Alto, Calif.) with 1-mm silica gel plates (Analtech) using a CH_2Cl_2/acetonitrile gradient. 2-Amino-4-(4-methoxyphenyl)thiazole (1a). ^{1}H NMR (400 MHz, DMSO-d_6) δ 7.71 (d, J $= 9$ Hz, 2H), 6.97 (bs, 2H), 6.90 (d, J $= 9$ Hz, 2H), 6.81 (s, 1H), 3.75 (s, 3H). (ESI-MS) m/z 207 (M$+1$). Calculated elemental analysis. C, 58.23; H, 4.89; N, 13.58; S, 15.54. Observed: C, 58.34; H, 5.01; N, 13.36; S, 15.39. All NMR spectra (400 MHz) were recorded on a Varian Instruments Gemini-400 spectrometer. The electrospray mass spectra (ESI-MS) were acquired on a Hewlett Packard 1100MSD spectrometer in the positive mode. Elemental analysis was done at Atlantic Microlab, Inc., Norcross, Ga.

5. Available from Ace Glassware Inc.

6. The rotating oven is available from Robbins Scientific.

7. Cleavage of the thiazoles from ArgoGel MB-CHO resin required longer cleavage times (4 h) and modest heating

(50°C). In addition, cleavage efficiency was enhanced when the resins were dried under vacuum before exposure to the TFA cleavage solution.

8. The ninhydrin test was performed according to a published procedure.[2] Potassium cyanide/pyridine (0.0002 M), phenol/ethanol (76% w/w), and ninhydrin/ethanol (0.28 M) were purchased from Perkin-Elmer.

DISCUSSION

The procedure described here illustrates a practical and efficient method for the solid-phase synthesis of 2-aminothiazoles, a useful structural element in medicinal chemistry. This structure has found application in drug development for the treatment of allergies,[3] hypertension,[4] inflammation,[5] schizophrenia,[6] and bacterial[7] and HIV[8] infections. The solid-phase route for the preparation of 2-aminothiazoles shown here can incorporate diverse functionality at each position of the molecule. A large number of the diversity reagents used in the synthesis are commercially available. In the procedure, resin-bound primary and secondary amines were converted to 1-substituted thioureas using fluorenylmethoxycarbonyl isothiocyanate (Fmoc-NCS).[9] The condensation of these immobilized thioureas with an α-bromoketone and subsequent acid cleavage produced the 2-aminothiazoles **1(a–e)** to **5(a–e)**. No linker was present in the cleaved material, and 2-aminothiazoles were formed in good purity and yields (54–96%) (Table 1.1).

The crude 2-aminothiazoles were dissolved in DMSO-d_6 (2 mL). A reference solution of *p*-dimethoxybenzene in DMSO-d_6 (2 M, 50 μL) was added to each of the samples, and proton NMR spectra were recorded. A 5 s delay was added between scans. The amount of 2-aminothiazole present was determined by a comparison of integral peak heights of the 2-aminothiazole and the reference compound.

TABLE 1.1. 2-Aminothiazole Yields

Entry	Product	Yield, %	Entry	Product	Yield, %
1	**1a**	86	14	**3d**	67
2	**1b**	91	15	**3e**	95
3	**1c**	96	16	**4a**	62
4	**1d**	57	17	**4b**	97
5	**1e**	87	18	**4c**	82
6	**2a**	54	19	**4d**	66
7	**2b**	68	20	**4e**	74
8	**2c**	69	21	**5a**	82
9	**2d**	57	22	**5b**	89
10	**2e**	70	23	**5c**	68
11	**3a**	61	24	**5d**	82
12	**3b**	91	25	**5e**	92
13	**3c**	87			

REFERENCES

1. Kearney, P. C.; Fernandez, M.; Flygare, J. A. *J. Org. Chem.* **1998**, *63*, 196.

2. Bunin, B. A. In, ed., *The Combinatorial Index*, Academic Press: San Diego, 1998, p. 214.

3. Hargrave, K. D.; Hess, F. K.; Oliver, J. T. *J. Med. Chem.* **1983**, *26*, 1158.

4. Patt, W. C.; Hamilton, H. W.; Taylor, M. D. et al. *J. Med. Chem.* **1992**, *35*, 2562.

5. Haviv, F.; Ratajczyk, J. D.; DeNet, R. W. et al. *J. Med. Chem.* **1988**, *31*, 1719; Clemence, F.; Martret, O. L.; Delevallee, F. et al. *J. Med. Chem.* **1988**, *31*, 1453.

6. Jaen, J. C.; Wise, L. D.; Caprathe, B. W. et al. *J. Med. Chem.* **1990**, *33*, 1453.

7. Tsuji, K.; Ishikawa, H. *Bioorg. Med. Chem. Lett.* **1994**, *4*, 1601.

8. Bell, F. W.; Cantrell, A. S.; Höberg, M. et al. *J. Med. Chem.* **1995**, *38*, 4929.

9. Kearney, P. C.; Fernandez, M.; Flygare, J. A. *Tetrahedron Lett.* **1998**, 2663.

CHAPTER TWO

SOLID-PHASE MANNICH REACTIONS OF A RESIN-IMMOBILIZED SECONDARY AMINE

Submitted by SCOTT L. DAX and JAMES J. McNALLY

Drug Discovery, The R. W. Johnson Pharmaceutical Research Institute, Welsh and McKean Roads, Spring House, PA, USA 19477

**Checked by BRIAN A. SIESEL,
THUY H. TRAN, and JENNIFER W. TAM**

Protein Design Labs, 34801 Campus Drive, Fremont, CA, USA 94555

LIBRARY SYNTHESIS ROUTE

a: 1.0 M aldehyde, 1.0 M alkyne, 7 Eq. 1,4-dimethylpiperazine, 1 Eq. Cu(I)Cl, dioxane, 100°C, 8 h.

b: TFA / H$_2$O (95:5).

BUILDING BLOCKS

benzaldehyde component:	X=	3-OCH$_3$	3-CH$_3$	2-Cl	4-Cl	3-OH	3-CN
		1	**2**	**3**	**4**	**5**	**6**
acetylene component:	R=	Ph	CH$_2$Ph	(CH$_2$)$_7$CH$_3$	Ph-2-Cl	C(CH$_3$)$_3$	
		A	**B**	**C**	**D**	**E**	

R–C≡CH

PROCEDURES

The piperazine trityl resin (Novabiochem, 1.55 mmol / g) was suspended in *N,N*-dimethylformamide (DMF) : dichloroethane (1 : 2 v/v) with gentle stirring to provide a uniform suspension of the resin (0.1 g resin / mL). Using a wide-bore pipette, a portion of this suspension (1 mL) was transferred to each reaction vessel to provide 0.10 g (0.155 mmol) of the resin-bound piperazine. The resin was washed twice with dioxane, and the solvent was drained. Copper(I) chloride (14–16 mg, ~0.15 mmol; note 1) was added to each reaction vessel followed by a solution of the appropriate acetylene in dioxane (2.0 M, 2.0 mL) and then 1,4-dimethylpiperazine (0.10 mL, 1.04 mmol; note 2), and the mixture was briefly agitated. A solution of the aldehyde component in dioxane (2.0 M, 2.0 mL) was added, and the reaction vessels were capped, agitated, and heated at 100°C for approximately 8 h. After cooling, the resins were filtered and washed sequentially with dioxane (1 × 2 mL), 10% piperidine in DMF (v/v) (4 × 2 mL), methanol (1 × 2 mL), 5% acetic acid in DMF (3 × 2 mL), methanol (3 × 2 mL), and finally methylene chloride (3 × 2 mL).

The resultant resins were separately treated with trifluoroacetic acid : water (95 : 5 v/v) (2 mL) for 5 min at ambient temperature and filtered. In each case, the filtrate was collected into a preweighed test tube. The resin was washed with an additional portion of trifluoroacetic acid : water mixture (2 mL of a 95 : 5 solution) and finally with dichloromethane (2 mL), and these washings were also collected. The combined filtrates were concentrated under a stream of nitrogen gas at 45°C to afford the crude product typically as a brown residue. This material was suspended in dichloromethane (2 mL), and the product was concentrated again under a stream of nitrogen. This procedure was repeated two more times to remove residual solvents. The resultant products were dried under vacuum overnight and the tubes were weighed to obtain the final yields of the products (Table 2.1). The products were typically obtained as brown glassy solids (note 3). A portion of the solid was removed and dissolved in methanol for HPLC and MS analysis (Table 2.2). The remainder of the product was dissolved in d_4-methanol or DMSO-d_6 for NMR analysis.

NOTES

1. Copper(I) chloride was ground to a fine powder with a mortar and pestle before use.

2. We have observed that 1,4-dimethylpiperazine is an innocuous additive that improves both the yield and crude purity of some

TABLE 2.1. Yield Ratio[a]

Component	A	B	C	D	E
1	83/83	72/85	81/88	94/88	100/80
2	86/80	114/83	79/86	85/86	76/77
3	79/84	73/86	74/89	85/90	84/80
4	83/84	77/86	77/89	88/89	83/80
5	88/81	93/83	81/86	114/86	83/78
6	75/82	75/84	76/88	74/87	79/79

[a] Isolated weight (mg)/theoretical weight (mg).

TABLE 2.2. Purity [a]

Component	A	B	C	D	E
1	95%	>95%	91%	>95%	>95%
2	94%	>95%	>95%	>95%	>95%
3	93%	>95%	>95%	>95%	>95%
4	93%	>95%	>95%	>95%	>95%
5	91%	83%	76%	83%	90%
6	73%	>95%	95%	27%	95%

[a] Determined by reverse-phase HPLC (acetonitrile–water gradient containing 0.1% TFA; 220 nM).

Mannich products. Accordingly, dimethylpiperazine was used in this array to provide uniform reaction conditions, although it is not needed for the formation and isolation of many Mannich adducts in this library.

3. Final products were isolated as solid glasses and typically contained minor amounts of residual trifluoroacetic acid, water, and/or dichloromethane.

DISCUSSION

To fully use the advantages afforded by multicomponent reaction systems in solid-phase organic synthesis, strategies in which each component is immobilized on the resin must be devised. In this way, individual components can be explored in terms of diversity without the restrictions imposed by immobilization. We have described solid-phase Mannich reactions[1] of a resin-bound alkyne (see chapter 5), and we show here that the diversity of products using this chemistry can be enhanced when a different component of the reaction system is immobilized. Specifically, a secondary amine, piperazine, is bound to a resin and then treated with various aldehydes and acetylenes in the presence of a copper(I) chloride catalyst to give a library of diverse Mannich adducts.[2]

A wide range of alkynes is tolerated, although in some cases substituted phenylacetylenes and acetylenes (e.g., 4-*t*-butylacetylene and 1-ethynylcyclohexene) gave polymeric material along with the desired products. In this study, we purposely limited the aldehyde component to a group of substituted benzaldehydes to provide a chromophore for analysis by HPLC using a UV detector, but in separate work we have shown that nonaromatic aldehydes (such as hexanal, paraformaldehyde, and cyclohexanecarboxaldehyde) also work well. A logical extension of this chemistry is immobilization of the aldehyde component and subsequent Mannich condensations to further enhance the diversity of compound libraries available through this chemistry. This work will be the subject of a future publication.

REFERENCES

1. Youngman, M. A.; Dax, S. L. *Tetrahedron Lett.* **1997**, *38*, 6347.
2. McNally, J. J.; Youngman, M. A.; Dax, S. L. *Tetrahedron Lett.* **1998**, *39*, 967.

CHAPTER THREE

SOLID-PHASE SYNTHESIS OF UREAS ON MICROTUBES

Submitted by HUI ZHUANG,[*] **EN-CHE YANG,**[†]
XIAO-YI XIAO, and A. W. CZARNIK[‡]

ChemRx / IRORI, Discovery Partners International
9640 Towne Centre Drive, San Diego, CA, USA 92121-1963

Checked by LEAH L. FRYE and RENEE ZINDELL

Boehringer, Ingelheim Pharmaceuticals, Inc., Research and
Development, 900 Ridgebury Road, Ridgefield, CT, USA 06877-0368

BUILDING BLOCKS

[*] To whom correspondence should be addressed. Tel: 858-546-3100; fax: 858-546-3083.
[†] Department of Chemistry, University of California, San Diego, CA.
[‡] Illumina, 9390 Towne Center Drive, La Jolla, CA 92121.

LIBRARY SYNTHESIS ROUTE

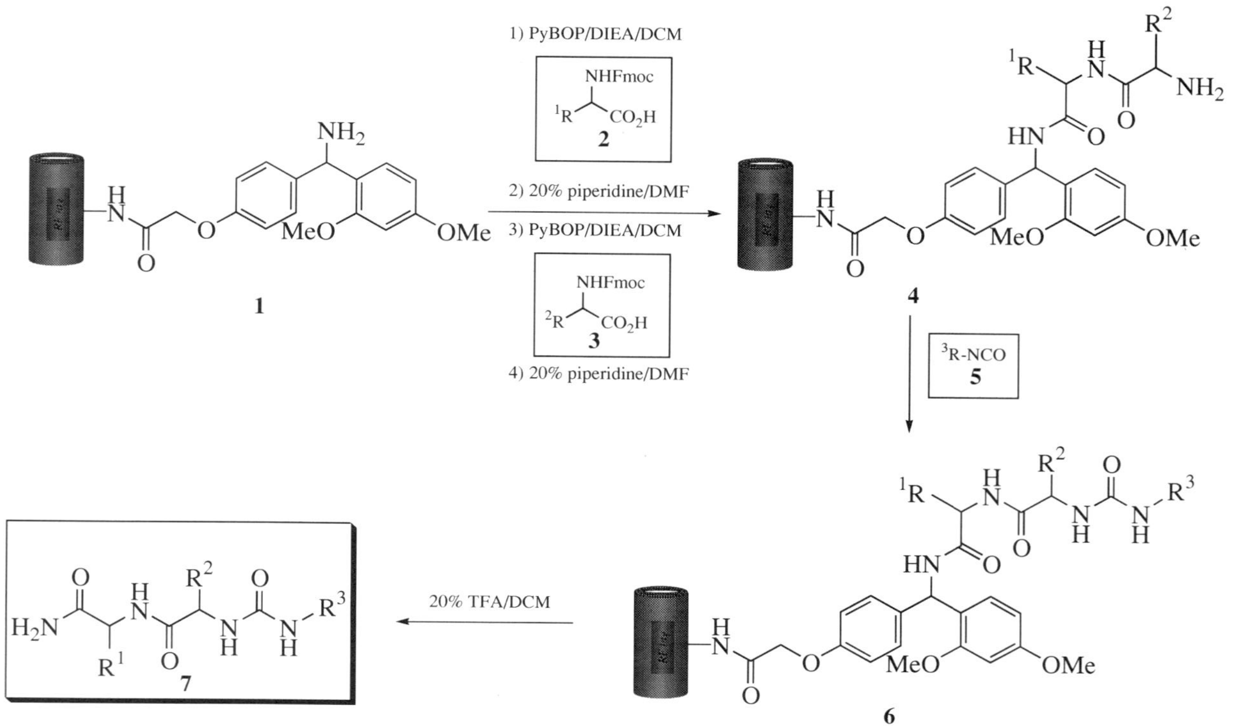

3 x 3 x 3 = 27

Scheme 1

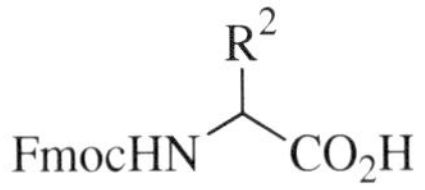

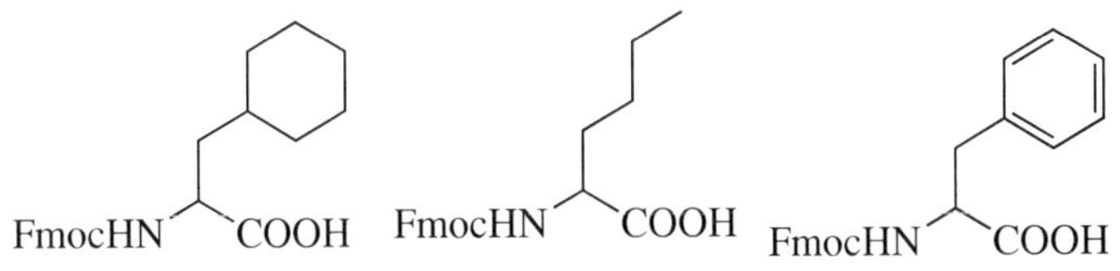

R^2

FmocHN CO_2H

FmocHN COOH   FmocHN COOH   FmocHN COOH

^{3}R-NCO

NCO

NCO

OMe

NCO

NO_2

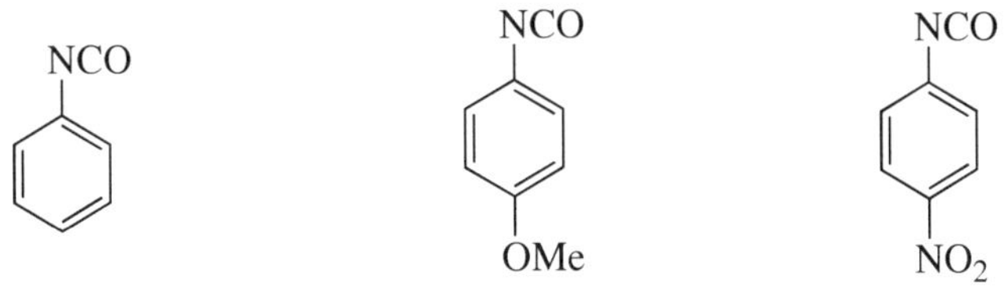

PROCEDURE

Loading Capacity Determination of Aminomethyl MicroTubes

Four aminomethyl MicroTubes (note 1) immersed in DCM (4 mL) were treated with Fmoc-Cl (0.104 g, 400 µmol; note 2) and DIEA (0.14 mL, 800 µmol). The reaction mixture was shaken (note 3) at room temperature for 2 h. After the supernatant was removed by aspiration, the MicroTubes were washed with MeOH, DCM, and ethyl ether (note 4) and dried under vacuum for 24 h. Each MicroTube was then treated with 2 mL of 20% piperidine in DMF at room temperature for 2 h. An aliquot (20 µL) of the solution was diluted to 1 mL with 20% piperidine in DMF. The loading was determined by measuring UV absorption of the solution at 301 nm ($\varepsilon_{max} = 7800$ M^{-1}cm^{-1}). A capacity loading of 46 µmol / MicroTube was obtained (note 5).

Rink Amide Linker Attachment

To 100 MicroTubes in 100 mL of DCM, the following were added sequentially: 4.96 g (9.2 mmol) Rink amide linker (note 6), 3.20 mL (18.4 mmol) DIEA, and 6.9 g (18.4 mmol) HATU (note 7). The reaction mixture was shaken at room temperature for 48 h. After the supernatant was removed by aspiration, the MicroTubes were washed sequentially with DMF, MeOH, and DCM for three cycles. The MicroTubes were dried under vacuum for 5 h after a final washing with ethyl ether.

Capping Conditions

After linker coupling, a positive Kaiser test is observed (note 8), which indicates a small amount of free NH$_2$. The free NH$_2$ can be capped with acetic anhydride. The above dried MicroTubes (**1**) were treated with a 100 mL solution of acetic anhydride (0.5 M) and DIEA (0.6 M) in DCM for 1–2 h. After the supernatant was removed by aspiration, the MicroTubes were washed three times with DMF, MeOH, and DCM and dried under vacuum for 3 h after a final washing with ethyl ether. The Kaiser test was negative.

De-Fmoc and Loading Measurement

A total of 40 MicroTubes were treated with 160 mL of 20% piperidine in DMF at room temperature for 60 min. An aliquot (40 µL) of the solution was diluted to 1 mL with 20% piperidine in DMF, and its UV absorption measured at 301 nm. After the supernate was removed by aspiration, the MicroTubes were washed with DMF, DCM, and MeOH three times. The MicroTubes were dried under vacuum for 24 h after a final washing with ethyl ether.

First Amino Acid Coupling

A total of 36 MicroTubes ($\sim$42 µmol/MicroTube) were sorted into three vials (note 9). MicroTubes in each vial were treated at room temperature with Fmoc-protected amino acids (**2**, 5.4 mmol, 10 equiv; note 10), DIEA (1.75 mL, 10.08 mmol, 20 Eq.), and HATU (1.91 g, 5.04 mmol, 10 Eq.) in DCM (24 mL) for 24 h. After the supernatant was removed by aspiration, the MicroTubes were then washed three times with DMF, DCM, MeOH, and DCM. The MicroTubes were dried under vacuum overnight. IR: 1657 cm^{-1} (CONHR; note 11).

Capping

The above dried MicroTubes were treated at room temperature with 60 mL of 0.6 M DIEA and 0.5 M acetic anhydride in DCM for 2 h. After the supernatant was removed by aspiration, the MicroTubes were washed three times with MeOH and DCM. The MicroTubes were dried under vacuum for 24 h after a final washing with ethyl ether.

De-Fmoc and Loading Measurement

A total of 33 MicroTubes were treated with 66 mL of 20% piperidine in DMF at room temperature for 2 h. An aliquot (40 µL) of the solution was diluted to 1 mL with 20% piperidine in DMF, and its UV absorption measured at 301 nm. After the supernatant was removed by aspiration, the MicroTubes were washed with MeOH and DCM three times. The MicroTubes were dried under vacuum for 24 h after a final washing with ethyl ether. (Loading: 38 µmol / MicroTube for Ala, 40 µmol / MicroTube for Phg, and 40 µmol / MicroTube for Leu.)

Second Amino Acid Coupling

A total of 33 MicroTubes were sorted and repooled into three vials, each containing 11 MicroTubes. Each vial was charged with one of three Fmoc-protected amino acids (**3**) (4.18 mmol, 10 Eq.; note 12), followed by addition of DIEA (9.95 mmol, 20 Eq.) and HATU (10 Eq.) in DCM (60 mL) for 48 h. After the supernatant was removed by aspiration, the MicroTubes were washed four times with MeOH and DCM and dried under vacuum overnight (note 13).

De-Fmoc and Loading Measurement

A total of 30 MicroTubes (**4**) were treated with 60 mL of 20% piperidine in DMF at room temperature for 2 h. An aliquot (20 µL) of the solution was diluted to 1 mL with 20% piperidine in DMF, and its UV absorption measured at 301 nm. (Loading: 38 µmol average.) After the supernate was removed by aspiration, the MicroTubes were washed with DMF, MeOH, and DCM three times. The MicroTubes were then dried under vacuum for 24 h after a final washing with ethyl ether.

Acylation: Preparation of Ureas

A total of 30 dipeptide MicroTubes (**4**) were resorted and pooled into three vials each containing 10 MicroTubes. Each vial was charged with one of three isocyanates (**5**) (3.8 mmol, 10 Eq.; note 14), and 20 mL of anhydrous DCM. The reactions were shaken at room temperature for 3 days. After the supernatant was removed by aspiration, the MicroTubes were washed with MeOH and

DCM five times. The MicroTubes were then dried under vacuum for 24 h after a final washing with ethyl ether (note 15).

Cleavage

A total of 27 MicroTubes (**6**) were sorted into 27 vials treated with 20% TFA in DCM (2 mL per vial) for 2 h. After the solution was collected, the MicroTubes were washed with DCM twice and the washing was combined with the original solution. The combined solutions were evaporated and the residue was dried under vacuum to yield 27 discrete compounds with purity ranging from 95 to 99%. The 27 final products were characterized by TLC, ^{1}H NMR, and MS (notes 16 and 17).

Description of Solid Support

Our library synthesis was carried out with a set of 27 tube-shaped solid phase synthesis support, called MicroTubes. These supports are prepared by radiation grafting of polystyrene ($\sim$350 µmol) onto polypropylene tubes, chemically functionalizing the polystyrene with aminomethyl groups to afford about 55 µmol of amine per tube, inserting a reusable Rf ID tag into each tube, and heat-sealing the tube ends to prevent loss of the tag. The chemical conversion of all 36 aminomethyl tubes was carried out simultaneously using standard procedures with rink amide linker, each with $\sim$46 µmol of available amine per tube.[1,2]

WASTE DISPOSAL INFORMATION

All toxic materials were disposed of in accordance with *Prudent Practices in the Laboratory* (Washington, D.C.: National Academy Press, 1995).

NOTES

1. Aminomethyl MicroTubes were obtained from the IRORI Division of Discovery Partners International. We just learned that the MicroTubes are no longer available, but that the chemistry has been shown to work well on loose resin by the Reviewer.

2. Fmoc-Cl was purchased from Advanced ChemTech.

3. We used an orbital shaker set at 200 rmp.

4. DCM, DIEA, MeOH, and ethyl ether were purchased from Aldrich and used as received.

5. UV measurements were performed on an HP 8452 diode array spectrophotometer.

6. Rink amide linker was purchased from Midwest Biotech.

7. HATU was purchased from PerSeptive BioSystem, GmbH.

8. The Kaiser test is a fast and sensitive color test capable of indicating whether greater than 99% of the terminal amino groups have reacted. This test is based on the reaction of ninhydrin reagent with small samples of amine resin or other solid support, such as MicroTubes. Three solutions are needed: (1) 500 mg ninhydrin in 10 mL ethanol, (2) 80 mg phenol in 20 mL ethanol, and (3) 2 mL 0.001 M solution of KCN diluted to 100 mL with pyridine. A small sample of the amine resin (1 to 2 mg) or a small piece of MicroTube surface (2 × 2 mm) was placed in a 12 × 75-mm test tube, and 2–3 drops of each of the three reagents were added. The test tube was kept in a heating block at 100°C for 5 min with occasional swirling. Upon observation, we found the beads or the piece of MicroTube surface to remain white and the solution yellow (negative test), indicating complete reaction. A dark blue color, which develops on the solid supports and in the solution, indicates a positive test.

9. Sorting was performed using IRORI's AccuTag-100 system. The 36 MicroTubes were sorted into three bottles, each containing 12 MicroTubes. The AccuTag-100 system uses electronic identification devices (radio frequency (Rf) ID tag) for encoding. With an Rf tag in each MicroTube, the tubes are initially scanned on the AccuTag-100 system, and the ID tag data are recorded via the *Synthesis Manager* software. Each Rf tag is associated with a compound in a chemical synthesis, thus allowing one to track the product through the process.

10. The first set of three amino acids are Fmoc-Ala-OH, Fmoc-Phg-OH, and Fmoc-Leu-OH, all purchased from Novabiochem.

11. One MicroTubes from each bottle were cleaved with 20% TFA in DCM for 2 h. After the solution was concentrated, the residues were dried and fully characterized with TLC, ^{1}H NMR, and MS to make sure that the reaction went to completion before the next step.

12. The second set of amino acids are Fmoc-Cha-OH, Fmoc-Nle-OH, and Fmoc-Phe-OH, all purchased from Novabiochem.

13. One MicroTubes from each bottle were cleaved with 20% TFA in DCM for 2 h. After the solvent was concentrated, the residues were dried and characterized by TLC, ^{1}H NMR, and MS before the next step.

14. The isocyanates used are phenylisocyanate, 4-nitrophenylisocyanate, and 4-methoxyphenylisocyanate, all purchased from Aldrich Chemical.

15. One MicroTubes from each bottle were cleaved with 20% TFA in DCM for 2 h. After the solvent was concentrated, the residues were dried and fully characterized by TLC, ^{1}H NMR, and MS before final cleavage.

16. ^{1}H NMR spectra were obtained on a 500 MHz Bruker NMR spectrometer with DMSO as the solvent and TMS as an internal standard, unless otherwise noted. Mass spectra were obtained on an Electrospray Spectrometer (M + Na).

17. The reviewer did the reaction on loose resin (because MicroTubes are no longer available) and washed it extensively in the isocyanate reaction with DMF to remove the unwanted symmetrical urea.

DISCUSSION

The urea functionality, a common structural motif in biologically active molecules,[3] is a nonhydrolyzable surrogate of an amide bond.[4] In our ongoing efforts to develop focused libraries of small molecules, there arose a need for the synthesis of unsymmetrical ureas. Although there are numerous classical methods known for the synthesis of symmetrical and unsymmetrical ureas,[5] the reaction of primary amines with isocyanates seems to be the method of choice for high-throughput synthesis.

Recently, Raju et al.[6] reported an attractive method for the preparation of unsymmetrical ureas on solid-phase resins, employing nitrophenylcarbamates as the key intermediates. They used this method to synthesize ureas derived from simple amines.[6] Here, we report that unsymmetrical ureas can be formed in high yield and purity using MicroTubes as the solid supports.

We prepared a combinatorial library that satisfied the following criteria: (1) the chemistry was general and applicable to a wide range of substrates; (2) the yields of all the transformations were high or the reactions were amenable to repetitive cycling under the reaction conditions to drive reactions to completion; (3) the reaction profiles were clean, minimizing the production of resin-bound impurities; and (4) the synthetic sequence minimized the number of chemical steps on solid support while maximizing the level of introduced diversity. Essentially, each synthetic transformation introduced a new point of diversity.

The preparation of ureas on MicroTubes is outlined in Scheme 1. Single coupling of Fmoc amino acids to the amines does not readily go to completion. Satisfactory results were obtained using double coupling with HATU.[7] The coupling step depends somewhat on the incoming amino acids.[8,9] Good yields were obtained with most amino acids, whereas relatively lower yields were obtained with hindered amino acids such as Val. Three MicroTube-bound intermediates from each step were verified by cleaving with 20% TFA in DCM, NMR, and MS analysis. After de-Fmoc, the intermediate **2** was then acylated using standard acylation procedures with Fmoc-Cha-OH, Fmoc-Nle-OH, and Fmoc-Phe-OH to provide **4**. Again, cleaving of three MicroTubes from each step under acidic conditions followed by spectroscopic analysis confirmed complete acylation. Once the dipeptide had been formed, the Fmoc protecting group was removed and urea formation was carried out by treatment with isocyanates.

For this library, we chose to use three types of isocyanates (neutral, electron rich, and electron deficient) to demonstrate the broad utility of the urea-formation reactions. Employing the above strategy and using the split-and-pool approach, we synthesized a 27-membered urea library with purities ranging from 95 to 99%. All the compounds prepared were characterized by ^{1}H NMR and mass spectroscopy. Acetonitrile can also be used as a substitute for DCM, but lower yields and product purities are generally observed. Attempts to use other protic solvents, such as isopropyl and ethyl alcohol, were unsuccessful. The best results were achieved when a chlorinated solvent (DCM) was used. The structure identity of all products was confirmed by ^{1}H NMR and MS spectroscopy. Expected molecular ions $(M + Na^+)$ were observed for all the products, and in all cases as the base peak. The compounds and yields are listed in Appendix 3.1.

Using this methodology, a library of thousands of compounds could be synthesized by using 20 amino acids and a few hundred isocyanates (about 300 are commercially available). As a follow-up to this 27-membered library, we did a reductive alkylation with aminomethyl MicroTubes first. Then identical procedures were applied all the way through to provide ureas that have four inputs. We had made nine compounds based on this route, and in all cases ~85% purity was achieved for each product.

In summary, we have described an efficient and facile solid-phase synthesis of substituted ureas starting from aminomethyl MicroTubes. The synthesis takes place under mild conditions. Taking into account the commercial availability of primary amines, this strategy can be ideally used for the synthesis of large combinatorial libraries.

REFERENCES

1. Li, R. S.; Xiao, X. Y.; Czarnik, A. W. *Tetrahedron Lett.* **1998**, *39*, 8681.

2. Zhao, C. F.; Shi, S.; Mir, D. et al. *J. Combinat. Chem.* **1999**, *1*, 91.

3. Majer, P.; Randad, R. S. *J. Org. Chem.* **1994**, *59*, 1937; Lefeber, D. J.; Liskamp, R. M. J. *Tetrahedron Lett.* **1997**, *38*, 5335.

4. Decieco, C. P.; Seng, J. L.; Kennedy, K. E. et al. *J. Bioorg. Med. Chem. Lett.* **1997**, *7*, 2331.

5. Katritzky, A. R.; Pleynet, D. P. M.; Yang, B. *J. Org. Chem.* **1997**, *62*, 4155; Xiao, X. Y.; Nug, K.; Chao, C.; Patel, D. V. *J. Org. Chem.* **1997**, *62*, 6968.

6. Raju, B.; Kassir, J. M; Kogan T. P. *J. Bioorg. Med. Chem. Lett.* **1998**, *8*, 3043.

7. Carpino, L. A.; Faham, E.; Minor, A.; Albericio, F. *J. Chem. Soc. Chem. Commun.* **1994**, 201.

8. Ostresh, J. M.; Winkle, J. H.; Hamashin, V. T.; Houghten, R. A. *Biopolymers*, **1994**, *34*, 1681.

9. Jay, M.; Ralph, A. R., *J. Org. Chem.* **1997**, *62*, 6090.

Appendix 3.1

N-(Phenylcarbamoyl)-L-Cha-L-Ala-NH$_2$ (1A4)

^{1}H NMR (DMSO) δ: 0.85–0.92 (m, 2H), 1.09–1.15 (m, 3H), 1.21 (d, J = 7.05 Hz, 3H, CH$_3$), 1.34–1.40 (m, 2H), 1.47–1.51 (m, 1H), 1.60–1.69 (m, 4H), 1.77–1.82 (m, 1H), 4.18–4.24 (m, 2H), 6.32 (d, J = 7.73 Hz, 1H, NH), 6.88 (t, J = 6.96 Hz, 1H), 6.98 (s, 1H, NH), 7.21 (t, J = 7.63 Hz, 3H), 7.36 (d, J = 7.80 Hz, 2H), 8.08 (d, J = 7.53 Hz, 1H, NH), and 8.60 (s, 1H, NH) ppm. MS/EI (C$_{19}$H$_{28}$N$_4$O$_3$) calculated: 360, observed: 383 ($+$Na$^+$).

N-(*p*-Methoxyphenylcarbamoyl)-L-Cha-L-Ala-NH$_2$ (1A5)

^{1}H NMR (DMSO) δ: 0.84–0.92 (m, 2H), 1.10–1.18 (m, 3H), 1.20 (d, J = 7.1 Hz, 3H, CH$_3$), 1.34–1.39 (m, 2H), 1.48–1.50 (m, 1H), 1.60–1.79 (m, 5H), 3.69 (s, 3H, OCH$_3$), 4.20 (m, 2H), 6.21 (d, J = 7.78 Hz, 1H, NH), 6.81 (d, J = 7.3 Hz, 2H), 6.98 (s, 1H, NH), 7.23 (brs, 1H, NH), 7.26 (d, J = 7.29 Hz, 2H), 8.06 (d, J = 7.66 Hz, 1H, NH), and 8.41 (s, 1H, NH) ppm. MS/EI (C$_{20}$H$_{30}$N$_4$O$_4$) calculated: 390; observed: 413 ($+$Na$^+$).

N-(*p*-Nitrophenylcarbamoyl)-L-Cha-L-Ala-NH$_2$ (1A6)

^{1}H NMR (DMSO) δ: 0.85 (m, 2H), 1.05–1.20 (m, 2H), 1.22 (d, J = 7.06 Hz, 3H, CH$_3$), 1.30–1.40 (m, 3H), 1.50–1.70 (m, 5H), 1.85 (m, 1H), 4.25 (m, 2H), 6.65 (d, J = 7.78 Hz, 1H, NH), 6.99 (brs, 1H, NH), 7.25 (brs, 1H, NH), 7.61 (d, J = 7.28 Hz, 2H), 8.14 (d, J = 9.43 Hz, 2H), 8.17 (d, J = 7.66 Hz, 1H, NH), and 9.41 (s, 1H, NH) ppm. MS/EI (C$_{19}$H$_{27}$N$_5$O$_5$) calculated: 405; observed: 428 ($+$Na$^+$).

N-(Phenylcarbamoyl)-L-Nle-L-Ala-NH$_2$ (1B4)

^{1}H NMR (DMSO) δ: 0.86 (t, 3H, CH$_3$), 1.21 (d, J = 7.06 Hz, 3H, CH$_3$), 1.27 (m, 6H), 4.17–4.24 (m, 2H), 6.36 (d, J = 7.87 Hz,

1H, NH), 6.89 (t, J = 7.65 Hz, 1H), 6.97 (brs, 1H, NH), 7.20 (t, J = 7.82 Hz, 2H), 7.25 (brs, 1H, NH), 7.35 (d, J = 8.05 Hz, 2H), 8.10 (d, J = 7.64 Hz, 1H, NH) and 8.64 (s, 1H, NH) ppm. MS/EI ($C_{16}H_{24}N_4O_3$) calculated: 320; observed: 343 ($+Na^+$).

N-(*p*-Methoxyphenylcarbamoyl)-L-Nle-L-Ala-NH₂ (1B5)

^{1}H NMR (DMSO) δ: 0.85 (t, J = 7.04 Hz, 3H, CH$_3$), 1.21 (d, J = 7.24 Hz, 3H, CH$_3$), 1.27 (m, 6H), 3.68 (s, 3H, OCH$_3$), 4.16–4.23 (m, 2H), 6.24 (d, J = 7.86 Hz, 1H, NH), 6.80 (d, J = 7.10 Hz, 2H), 6.97 (brs, 1H, NH), 7.25–7.27 (m, 3H), 8.07 (d, 1H, NH), and 8.44 (s, 1H, NH) ppm. MS/EI ($C_{17}H_{26}N_4O_4$) calculated: 350, observed: 373 ($+Na^+$).

N-(*p*-Nitrophenylcarbamoyl)-L-Nle-L-Ala-NH₂ (1B6)

^{1}H NMR (DMSO) δ: 0.86 (t, J = 6.83 Hz, 3H, CH$_3$), 1.22 (d, J = 6.99 Hz, 3H, CH$_3$), 1.28 (m, 4H), 1.52–1.55 (m, 1H), 1.65–1.68 (m, 1H), 4.22–4.26 (m, 2H), 6.68 (d, J = 7.96 Hz, 1H, NH), 6.98 (brs, 1H, NH), 7.27 (brs, 1H, NH), 7.59–7.61 (d, J = 9.03 Hz, 2H), 8.14 (d, J = 8.93 Hz, 2H), 8.18 (d, J = 7.68 Hz, 1H, NH), and 9.44 (s, 1H, NH) ppm. MS/EI ($C_{16}H_{23}N_5O_5$) calculated: 365, observed: 388 ($+Na^+$).

N-(Phenylcarbamoyl)-L-Phe-L-Ala-NH₂ (1C4)

^{1}H NMR (DMSO) δ: 1.22 (d, J = 6.97 Hz, 3H, CH$_3$), 2.72–2.88 (dd, J$_1$ = 8.37 Hz, J$_2$ = 8.41 Hz, 1H), 3.03–3.07 (dd, J$_1$ = 4.63 Hz, J$_2$ = 4.72 Hz, 1H), 4.22–4.25 (m, 1H), 4.49–4.51 (m, 1H), 6.27 (d, J = 7.95 Hz, 1H, NH), 6.87 (t, J = 6.92 Hz, 1H), 7.02 (brs, 1H, NH), 7.18–7.32 (m, 10H), 8.21 (d, J = 7.67 Hz, 1H, NH), and 8.67 (s, 1H, NH) ppm. MS/EI ($C_{19}H_{22}N_4O_3$) calculated: 354; observed: 377 ($+Na^+$).

N-(*p*-Methoxyphenylcarbamoyl)-L-Phe-L-Ala-NH$_2$ (1C5)

^{1}H NMR (DMSO) δ: 1.22 (d, J = 7.14 Hz, 3H, CH$_3$), 2.78–2.82 (dd, J$_1$ = 8.19 Hz, J$_2$ = 8.21 Hz, 1H), 3.02–3.05 (dd, J$_1$ = 4.59 Hz, J$_2$ = 4.63 Hz, 1H), 3.67 (s, 3H, OCH$_3$), 4.21–4.24 (m, 1H), 4.50 (m, 1H), 6.16 (d, J = 7.96 Hz, 1H, NH), 6.78 (d, J = 8.92 Hz, 2H), 7.02 (brs, 1H, NH), 7.17–7.28 (m, 7H), 8.18 (d, 1H, NH), and 8.48 (s, 1H, NH) ppm. MS/EI (C$_{20}$H$_{24}$N$_4$O$_4$) calculated: 384; observed: 407 (+Na$^+$).

N-(*p*-Nitrophenylcarbamoyl)-L-Phe-L-Ala-NH$_2$ (1C6)

^{1}H NMR (DMSO) δ: 1.25 (d, J = 7.38 Hz, 3H, CH$_3$), 2.82–2.86 (dd, J$_1$ = 8.11 Hz, J$_2$ = 8.13 Hz, 1H), 3.06–3.10 (dd, J$_1$ = 4.58 Hz, J$_2$ = 4.73 Hz, 1H), 4.23–4.31 (m, 1H), 4.54–4.59 (m, 1H), 6.57–6.58 (d, J = 8.04 Hz, 1H, NH), 7.03 (brs, 1H, NH), 7.17–7.28 (m, 6H), 7.56 (s, 2H), 8.13 (d, J = 9.16 Hz, 2H), 8.31 (d, J = 7.67 Hz, 1H, NH) and 9.46 (s, 1H, NH) ppm. MS/EI (C$_{19}$H$_{21}$N$_5$O$_5$) calculated: 399; observed: 422 (+Na$^+$).

N-(Phenylcarbamoyl)-L-Cha-L-Phg-NH$_2$ (2A4)

^{1}H NMR (DMSO) δ: 0.83–0.92 (m, 2H), 1.07–1.22 (m, 3H), 1.33–1.79 (m, 8H), 4.37–4.41 (m, 1H), 5.39–5.42 (m, 1H), 6.37 (d, J = 8.23 Hz, 1H, NH), 6.87–6.91 (m, 1H, NH), 7.16–7.43 (m, 10H), 7.68 (d, J = 7.97 Hz, 1H, NH), 8.52 (d, J = 10.41 Hz, 1H, NH), and 8.59 (d, J = 8.18 Hz, 1H, NH) ppm. MS/EI (C$_{24}$H$_{30}$N$_4$O$_3$) calculated: 422; observed: 445 (+Na$^+$).

N-(*p*-Methoxyphenylcarbamoyl)-L-Cha-L-Phg-NH$_2$ (2A5)

^{1}H NMR (DMSO) δ: 0.83–0.91 (m, 2H), 1.07–1.43 (m, 5H), 1.47–1.78 (m, 6H), 3.69 (s, 3H, OCH$_3$), 4.34–4.38 (m, 1H), 5.38–

5.41 (m, 1H), 6.26 (d, J = 8.16 Hz, 1H, NH), 6.80–6.82 (m, 2H), 7.21–7.34 (m, 5H), 7.40–7.43 (m, 2H), 7.68 (s, 1H, NH), 8.39 (d, J = 10.05 Hz, 1H, NH) and 8.48 (d, J = 8.06 Hz, 1H, NH) ppm. MS/EI ($C_{25}H_{32}N_4O_4$) calculated: 452, observed: 475 ($+Na^+$).

N-(*p*-Nitrophenylcarbamoyl)-L-Cha-L-Phg-NH$_2$ (2A6)

^{1}H NMR (DMSO) δ: 0.85–0.92 (m, 2H), 1.07–1.20 (m, 3H), 1.34–1.48 (m, 2H), 1.50–1.89 (m, 6H), 4.50 (m, 1H), 5.45 (m, 1H), 6.69 (d, J = 8.31 Hz, 1H, NH), 7.18 (brs, 1H, NH), 7.25–7.44 (m, 6H), 7.59 (d, J = 9.45 Hz, 2H), 8.14 (m, 2H), 8.63 (d, J = 8.13 Hz, 1H, NH), and 9.39 (d, 1H, NH) ppm. MS/EI ($C_{24}H_{29}N_5O_5$) calculated: 467, observed: 490 ($+Na^+$).

N-(Phenylcarbamoyl)-L-Nle-L-Phg-NH$_2$ (2B4)

^{1}H NMR (DMSO) δ: 0.7–0.8 (tt, $J_1 = 7.38$ Hz, $J_2 = 6.79$ Hz, 3H), 1.29–1.41 (m, 4H), 1.52–1.70 (m, 2H), 4.45 (m, 1H), 5.45 (m, 1H), 6.45 (d, J = 8.33 Hz, 1H, NH), 6.9 (m, 1H, NH), 7.20–7.44 (m, 9H), 7.70 (m, 1H), 8.61 (d, J = 7.98 Hz, 1H, NH), and 8.65 (s, 1H) ppm. MS/EI ($C_{21}H_{26}N_4O_3$) calculated: 382; observed: 405 ($+Na^+$).

N-(*p*-Methoxyphenylcarbamoyl)-L-Nle-L-Phg-NH$_2$ (2B5)

^{1}H NMR (DMSO) δ: 0.7–0.8 (tt, $J_1 = 7.01$ Hz, $J_2 = 6.87$ Hz, 3H, CH$_3$), 1.27 (m, 4H), 1.60 (m, 2H), 3.68 (s, 3H, OCH$_3$), 4.35 (m, 1H), 5.45 (m, 1H), 6.35 (d, J = 8.10 Hz, 1H, NH), 6.80 (m, 2H), 7.25 (d, 1H, NH), 7.26–7.32 (m, 5H), 7.42 (m, 2H), 7.70 (s, 1H, NH), 8.42 (s, 1H, NH), and 8.55 (d, J = 8.04 Hz, 1H, NH) ppm. MS/EI ($C_{22}H_{28}N_4O_4$) calculated: 412; observed: 435 ($+Na^+$).

N-(*p*-Nitrophenylcarbamoyl)-L-Nle-L-Phg-NH$_2$ (2B6)

^{1}H NMR (DMSO) δ: 0.86–0.90 (tt, J$_1$ = 7.44 Hz, J$_2$ = 7.36 Hz, 3H, CH$_3$), 1.29 (m, 4H), 1.55–1.75 (m, 2H), 4.50 (m, 1H), 5.45 (m, 1H), 6.75 (d, J = 8.10 Hz, 1H, NH), 7.20 (s, 1H, NH), 7.32–7.45 (m, 5H), 7.59 (m, 2H), 7.75 (s, 1H, NH), 8.13 (m, 2H), 8.65 (d, J = 8.10 Hz, 1H, NH), and 9.43 (s, 1H, NH) ppm. MS/EI (C$_{21}$H$_{25}$N$_2$O$_5$) calculated: 427; observed 450 (+Na$^+$).

N-(Phenylcarbamoyl)-L-Phe-L-Phg-NH$_2$ (2C4)

^{1}H NMR (DMSO) δ: 2.85 (m, 1H), 3.05 (m, 1H), 4.70 (m, 1H), 5.45 (m, 1H), 6.35 (m, 1H, NH), 6.95 (m, 1H), 7.19–7.32 (m, 15H), 7.45 (s, 1H, NH), and 8.70 (s, 1H, NH) ppm. MS/EI (C$_{24}$H$_{24}$N$_4$O$_3$) calculated: 416; observed: 439 (+Na$^+$).

N-(*p*-Methoxyphenylcarbamoyl)-L-Phe-L-Phg-NH$_2$ (2C5)

^{1}H NMR (DMSO) δ: 2.82–2.86 (dd, J$_1$ = 8.31 Hz, J$_2$ = 8.35 Hz, 1H), 3.02–3.06 (dd, J$_1$ = 4.69 Hz, J$_2$ = 4.73 Hz, 1H), 3.67 (s, 3H, OCH$_3$), 4.65 (m, 1H), 5.45 (m, 1H), 6.25 (m, 1H, NH), 6.78 (d, 2H), 7.21–7.26 (m, 12H), 7.43 (d, 2H), 7.75 (d, 1H, NH), and 8.50 (s, 1H, NH) ppm. MS/EI (C$_{25}$H$_{26}$N$_4$O$_4$) calculated: 446; observed: 469 (+Na$^+$).

N-(*p*-Nitrophenylcarbamoyl)-L-Phe-L-Phg-NH$_2$ (2C6)

^{1}H NMR (DMSO) δ: 2.87–2.91 (dd, J$_1$ = 8.02 Hz, J$_2$ = 8.06 Hz, 1H), 3.07–3.11 (dd, J$_1$ = 4.54 Hz, J$_2$ = 4.61 Hz, 1H), 4.75 (m, 1H), 5.45 (m, 1H), 6.60 (d, J = 8.19 Hz, 1H, NH), 7.19–7.40 (m, 10H), 7.50 (d, 2H), 7.55 (m, 2H), 8.11 (m, 2H), 8.79 (d, J = 8.16 Hz, 1H, NH), and 9.45 (s, 1H, NH) ppm. MS/EI (C$_{24}$H$_{23}$N$_5$O$_5$) calculated: 461; observed: 484 (+Na$^+$).

N-(Phenylcarbamoyl)-L-Cha-L-Leu-NH$_2$ (3A4)

^{1}H NMR (DMSO) δ: 0.82 (d, J = 6.70 Hz, 3H, CH$_3$), 0.86 (d, J = 6.64 Hz, 3H, CH$_3$), 1.10–1.20 (m, 3H), 1.30–1.51 (m, 6H), 1.62–1.69 (m, 7H), 4.25 (m, 2H), 6.35 (d, J = 7.86 Hz, 1H, NH), 6.87–6.97 (m, 1H, NH), 7.22 (t, J = 8.29 Hz, 3H), 7.35 (d, J = 8.25 Hz, 2H), 8.05 (d, J = 8.31 Hz, 1H), and 8.60 (s, 1H) ppm. MS/EI (C$_{22}$H$_{34}$N$_4$O$_3$) calculated: 402; observed: 425 (+Na$^+$).

N-(*p*-Methoxyphenylcarbamoyl)-L-Cha-L-Leu-NH$_2$ (3A5)

^{1}H NMR (DMSO) δ: 0.82 (d, J = 8.85 Hz, 3H, CH$_3$), 0.87 (d, J = 8.98 Hz, 3H, CH$_3$), 1.10–1.70 (m, 16H), 3.69 (s, 3H, OCH$_3$), 4.25 (m, 2H), 6.20 (m, 1H, NH), 6.81 (d, J = 8.85 Hz, 2H), 7.0 (s, 1H, NH), 7.23 (s, 1H, NH), 7.27 (d, J = 7.09 Hz, 2H), 8.0 (d, J = 8.56 Hz, 1H, NH), and 8.42 (s, 1H, NH) ppm. MS/EI (C$_{23}$H$_{36}$N$_4$O$_4$) calculated: 432; observed: 455 (+Na$^+$).

N-(*p*-Nitrophenylcarbamoyl)-L-Cha-L-leu-NH$_2$ (3A6)

^{1}H NMR (DMSO) δ: 0.83 (d, J = 6.25 Hz, 3H, CH$_3$), 0.87 (d, J = 6.66 Hz, 3H, CH$_3$), 1.15 (m, 4H), 1.31–1.60 (m, 6H), 1.65–1.79 (m, 6H), 4.29 (m, 2H), 6.65 (d, 1H, NH), 7.0 (s, 1H, NH), 7.25 (s, 1H, NH), 7.60 (d, J = 8.91 Hz, 2H), 8.09 (d, J = 9.10 Hz, 1H, NH), 8.14 (d, 2H), and 9.41 (s, 1H, NH) ppm. MS/EI (C$_{22}$H$_{33}$N$_5$O$_5$) calculated: 447; observed: 470 (+Na$^+$).

N-(Phenylcarbamoyl)-L-Nle-L-Leu-NH$_2$ (3B4)

^{1}H NMR (DMSO) δ: 0.83 (d, J = 6.64 Hz, 3H, CH$_3$), 0.86 (t, 3H, CH$_3$), 0.89 (d, J = 6.72 Hz, 3H, CH$_3$), 1.25–1.28 (m, 4H), 1.45–1.60 (m, 5H), 4.25 (m, 2H), 6.36 (d, J = 7.22 Hz, 1H, NH), 6.87–6.97 (m, 1H, NH), 7.21 (t, J = 8.25 Hz, 2H), 7.27 (s, 1H,

NH), 7.36 (d, J = 7.57 Hz, 2H), 8.04 (d, J = 8.44 Hz, 1H, NH), and 8.65 (s, 1H, NH) ppm. MS/EI ($C_{19}H_{30}N_4O_4$) calculated: 362; observed: 385 ($+Na^+$).

N-(*p*-Methoxyphenylcarbamoyl)-L-Nle-L-Leu-NH$_2$ (3B5)

^{1}H NMR (DMSO) δ: 0.83 (d, J = 6.55 Hz, 3H, CH$_3$), 0.87 (d, J = 6.68 Hz, 3H, CH$_3$), 0.89 (m, 3H, CH$_3$), 1.26–1.65 (m, 9H), 3.69 (s, 3H, OCH$_3$), 4.09–4.30 (m, 2H), 6.3 (d, J = 7.69 Hz, 1H, NH), 6.80 (d, J = 7.07 Hz, 2H), 6.82–6.97 (m, 1H, NH), 7.25 (d, J = 7.15 Hz, 2H), 7.35 (d, J = 8.83 Hz, 1H, NH), 8.0 (s, 1H, NH) and 8.45 (s, 1H, NH) ppm. MS/EI ($C_{20}H_{32}N_4O_4$) calculated: 392; observed: 415 ($+Na^+$).

N-(*p*-Nitrophenylcarbamoyl)-L-Nle-L-Leu-NH$_2$ (3B6)

^{1}H NMR (DMSO) δ: 0.83 (d, J = 6.62 Hz, 3H, CH$_3$), 0.85 (t, 3H, CH$_3$), 0.88 (d, J = 6.37 Hz, 3H, CH$_3$), 1.27 (m, 3H), 1.4–1.60 (m, 6H), 4.25 (m, 2H), 6.65 (d, J = 7.82 Hz, 1H, NH), 6.97 (s, 1H, NH), 7.29 (s, 1H, NH), 7.59 (d, J = 8.93 Hz, 2H), 8.14 (d, J = 8.99 Hz, 2H), and 9.47 (s, 1H, NH) ppm. MS/EI ($C_{19}H_{29}N_5O_5$) calculated: 407; observed: 430 ($+Na^+$).

N-(Phenylcarbamoyl)-L-Phe-L-Leu-NH$_2$ (3C4)

^{1}H NMR (DMSO) δ: 0.83 (d, J = 6.69 Hz, 3H, CH$_3$), 0.87 (d, J = 6.74 Hz, 3H, CH$_3$), 1.45–1.49 (m, 2H), 1.55–1.59 (m, 1H), 2.81–2.85 (dd, J$_1$ = 7.88 Hz, J$_2$ = 7.89 Hz, 1H), 3.02–3.05 (dd, J$_1$ = 4.81 Hz, J$_2$ = 4.85 Hz, 1H), 4.24–4.27 (m, 1H), 4.50–4.52 (m, 1H), 6.27 (d, J = 7.83 Hz, 1H, NH), 6.88–7.00 (m, 1H, NH), 7.17–7.27 (m, 9H), 7.32 (d, J = 7.73 Hz, 1H, NH), 8.14 (d, J = 8.05 Hz, 1H, NH), and 8.68 (s, 1H, NH) ppm. MS/EI ($C_{22}H_{28}N_4O_3$) calculated: 396; observed: 419 ($+Na^+$).

N-(*p*-Methoxyphenylcarbamoyl)-L-Phe-L-Leu-NH$_2$ (3C5)

^{1}H NMR (DMSO) δ: 0.83 (d, J = 6.60 Hz, 3H, CH$_3$), 0.87 (d, J = 6.72 Hz, 3H, CH$_3$), 1.45–1.49 (m, 2H), 1.56–1.57 (m, 1H), 2.80–2.85 (dd, J$_1$ = 7.91 Hz, J$_2$ = 7.93 Hz, 1H), 3.00–3.04 (dd, J$_1$ = 4.81 Hz, J$_2$ = 4.85 Hz, 1H), 3.68 (s, 3H, OCH$_3$), 4.25–4.27 (m, 1H), 4.49–4.50 (m, 1H), 6.16 (d, J = 7.44 Hz, 1H, NH), 6.78 (d, J = 8.93 Hz, 2H), 7.00 (s, 1H, NH), 7.17–7.27 (m, 8H), 8.11 (d, J = 8.33 Hz, NH), and 8.50 (s, 1H, NH) ppm. MS/EI (C$_{23}$H$_{30}$N$_4$O$_4$) calculated: 426; observed: 449 (+Na$^+$).

N-(*p*-Nitrophenylcarbamoyl)-L-Phe-L-Leu-NH$_2$ (3C6)

^{1}H NMR (DMSO) δ: 0.84 (d, J = 6.69 Hz, 3H, CH$_3$), 0.88 (d, J = 6.44 Hz, 3H, CH$_3$), 1.46–1.49 (m, 2H), 1.57–1.60 (m, 1H), 2.84–2.89 (dd, J$_1$ = 7.71 Hz, J$_2$ = 8.32 Hz, 1H), 3.05–3.09 (dd, J$_1$ = 4.70 Hz, J$_2$ = 4.79 Hz, 1H), 6.56 (d, J = 7.96 Hz, 1H, NH), 7.02 (s, 1H, NH), 7.17–7.27 (m, 6H), 7.57 (d, J = 9.50 Hz, 2H), 8.12 (d, J = 9.11 Hz, 2H), 8.25 (d, J = 8.51 Hz, 1H, NH), and 9.48 (s, 1H, NH) ppm. MS/EI (C$_{22}$H$_{27}$N$_5$O$_5$) calculated: 441; observed: 464 (+Na$^+$).

Appendix 3.2

Analytical Data of the Urea Library

Entry	Chemical Formula	Code	Structure	[M+Na]$^+$ (exact mass ^{a})	Purity ^{b}	Quantity (percent) ^{c}
1	C$_{19}$H$_{28}$N$_4$O$_3$	1A4	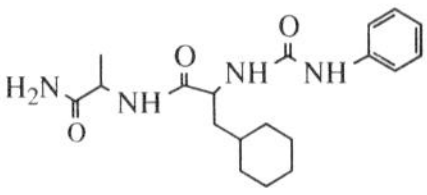	383(360)	high	99
2	C$_{20}$H$_{30}$N$_4$O$_4$	1A5	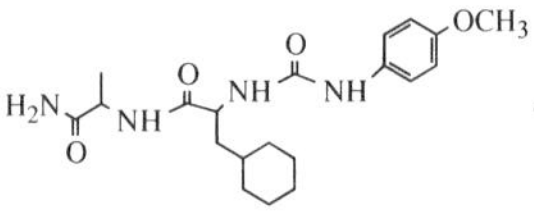	413(390)	high	96

Appendix 3.2 (*Continued*)

Entry	Chemical Formula	Code	Structure	$[M+Na]^+$ (exact mass[a])	Purity[b]	Quantity (percent)[c]
3	$C_{19}H_{27}N_5O_5$	1A6	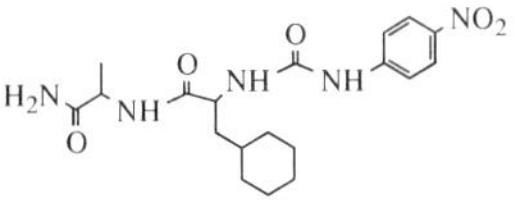	428(405)	high	89
4	$C_{16}H_{24}N_4O_3$	1B4	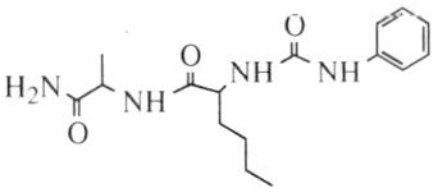	343(320)	high	90
5	$C_{17}H_{26}N_4O_4$	1B5	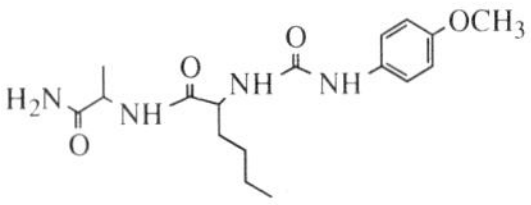	373(350)	high	87
6	$C_{16}H_{23}N_5O_5$	1B6	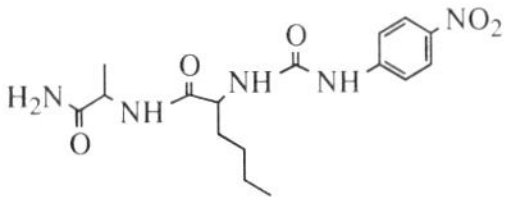	388(365)	high	99
7	$C_{19}H_{22}N_4O_3$	1C4	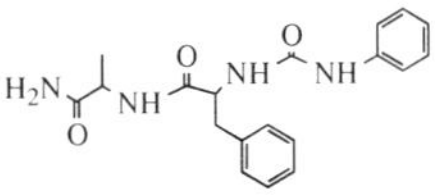	377(354)	high	86
8	$C_{20}H_{24}N_4O_4$	1C5	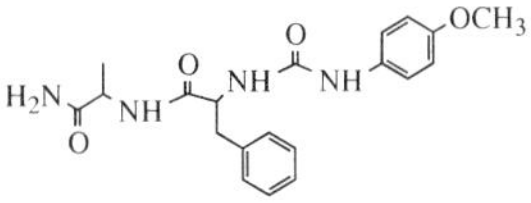	407(384)	high	96
9	$C_{19}H_{21}N_5O_5$	1C6	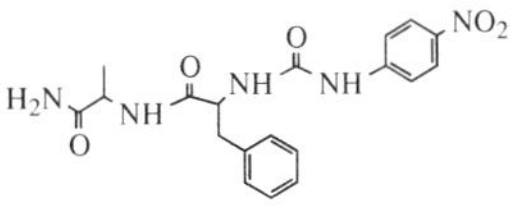	422(399)	high	99
10	$C_{24}H_{30}N_4O_3$	2A4	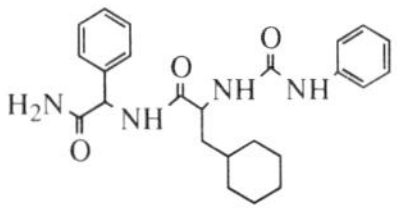	445(422)	high	99
11	$C_{25}H_{32}N_4O_4$	2A5	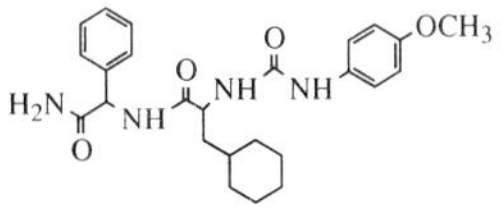	475(452)	high	97
12	$C_{24}H_{29}N_5O_5$	2A6	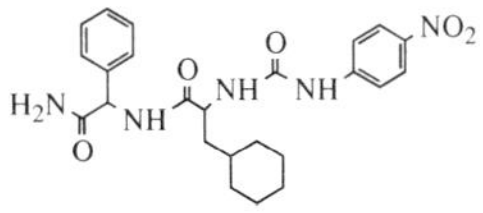	490(467)	high	80

Appendix 3.2 (*Continued*)

Entry	Chemical Formula	Code	Structure	$[M+Na]^+$ (exact mass[a])	Purity[b]	Quantity (percent)[c]
13	$C_{21}H_{26}N_4O_3$	2B4	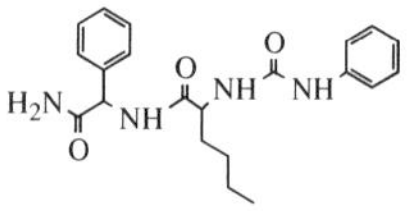	405(382)	high	82
14	$C_{22}H_{28}N_4O_4$	2B5	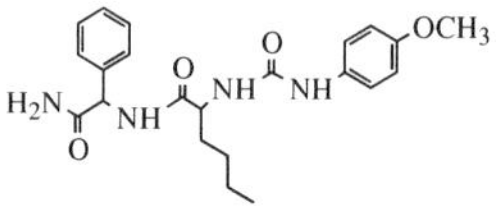	435(412)	high	99
15	$C_{21}H_{25}N_5O_5$	2B6	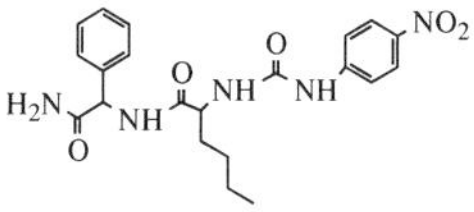	450(427)	high	92
16	$C_{24}H_{24}N_4O_3$	2C4	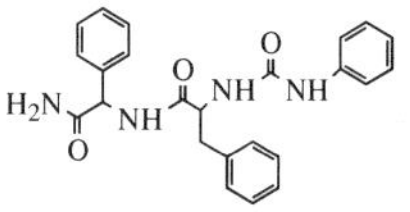	439(416)	high	96
17	$C_{25}H_{26}N_4O_4$	2C5	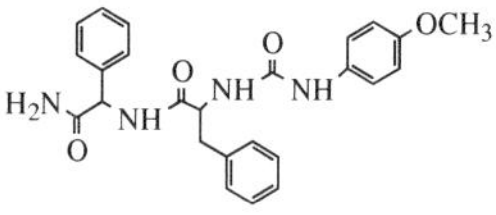	469(446)	high	94
18	$C_{24}H_{23}N_5O_5$	2C6	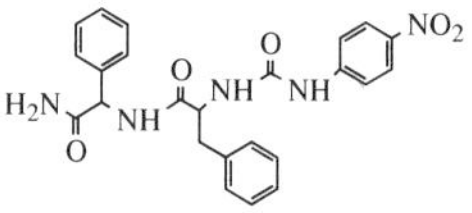	484(461)	high	96
19	$C_{22}H_{34}N_4O_3$	2A4	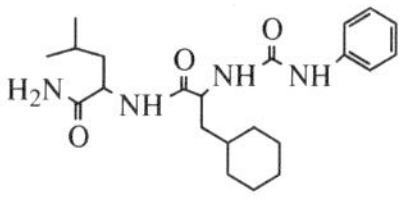	425(402)	high	97
20	$C_{23}H_{36}N_4O_4$	3A5	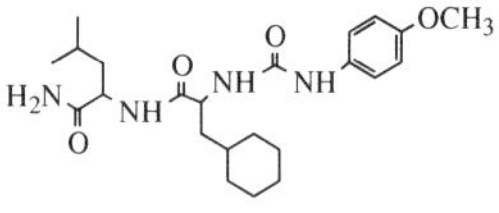	455(432)	high	89
21	$C_{22}H_{33}N_5O_5$	3A6	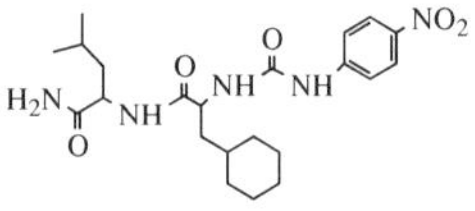	470(447)	high	80
22	$C_{19}H_{30}N_4O_3$	3B4	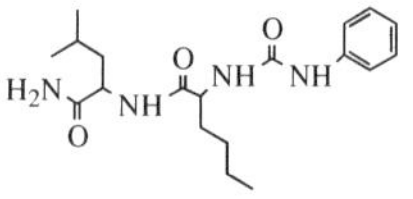	385(362)	high	99

Appendix 3.2 (*Continued*)

Entry	Chemical Formula	Code	Structure	[M+Na]$^+$ (exact mass^a)	Purity^b	Quantity (percent)^c
23	$C_{20}H_{32}N_4O_4$	3B5	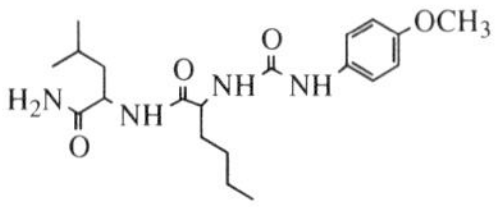	415(392)	high	99
24	$C_{19}H_{29}N_5O_5$	3B6	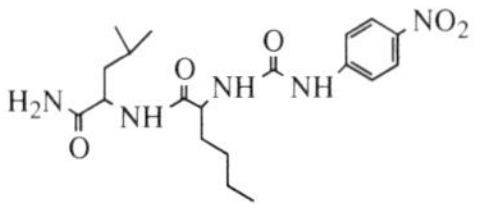	430(407)	high	90
25	$C_{22}H_{28}N_4O_3$	3C4	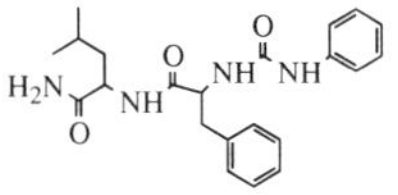	419(396)	high	98
26	$C_{23}H_{30}N_4O_4$	3C5	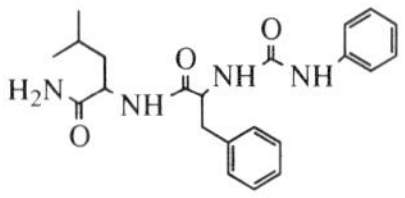	449(426)	high	80
27	$C_{22}H_{27}N_5O_5$	3C6	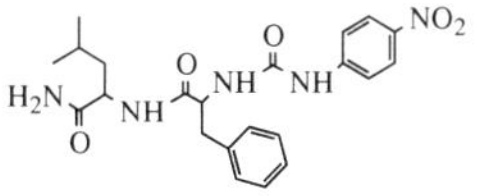	464(441)	high	98

^{a} Data were obtained by electron spray mass spectrometry analysis.
^{b} Estimated by ^{1}H NMR analysis (DMSO). *High*, >80% pure; *medium*, 50–80% pure; *low*, <50% pure.
^{c} Estimated by ^{1}H NMR analysis with an internal standard (TMS).

Appendix 3.3

Solid-Phase Synthesis of Ureas on MicroTubes

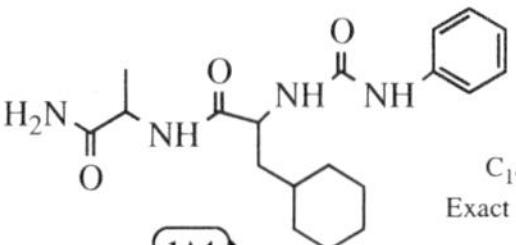 $C_{19}H_{28}N_4O_3$ Exact Mass: 360.22 **1A4**	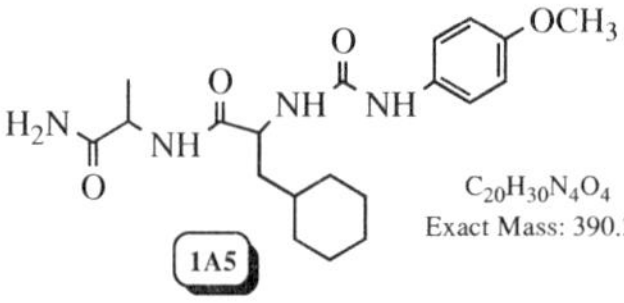 $C_{20}H_{30}N_4O_4$ Exact Mass: 390.23 **1A5**
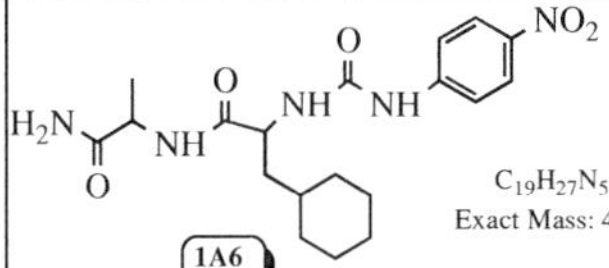 $C_{19}H_{27}N_5O_5$ Exact Mass: 405.20 **1A6**	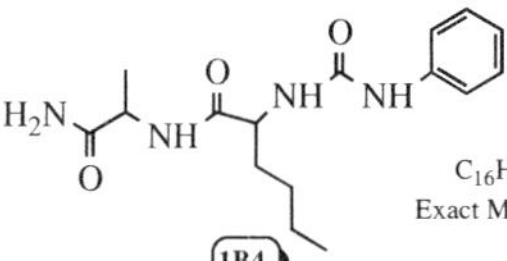 $C_{16}H_{24}N_4O_3$ Exact Mass: 320.18 **1B4**
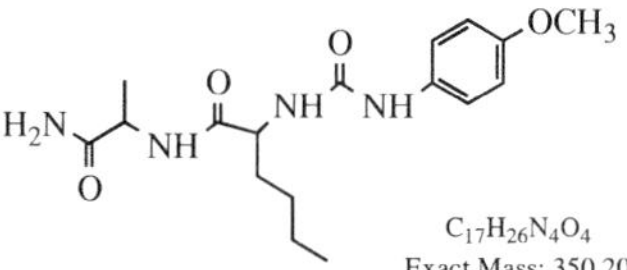 $C_{17}H_{26}N_4O_4$ Exact Mass: 350.20 **1B5**	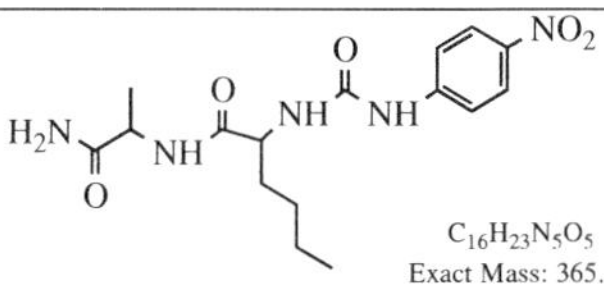 $C_{16}H_{23}N_5O_5$ Exact Mass: 365.17 **1B6**
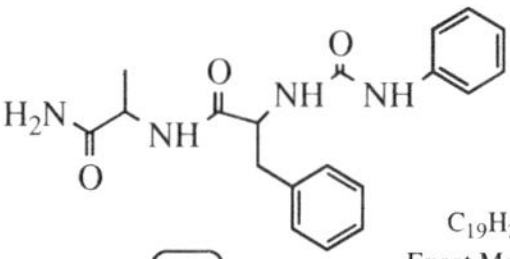 $C_{19}H_{22}N_4O_3$ Exact Mass: 354.17 **1C4**	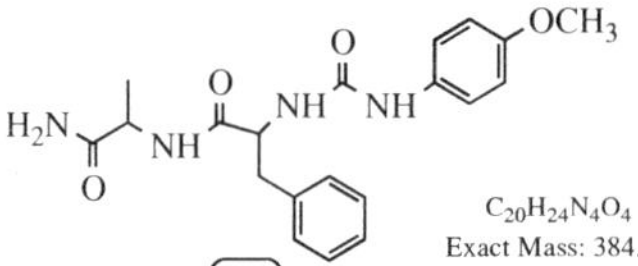 $C_{20}H_{24}N_4O_4$ Exact Mass: 384.18 **1C5**
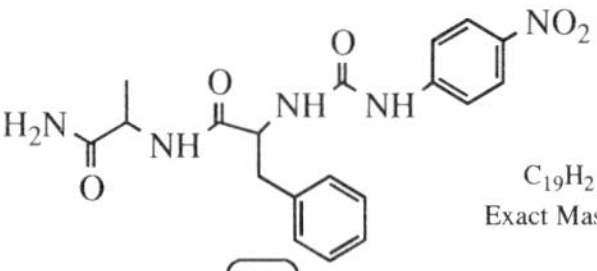 $C_{19}H_{21}N_5O_5$ Exact Mass: 399.15 **1C6**	

Appendix 3.3 (*Continued*)

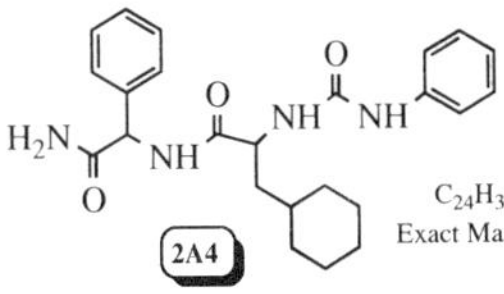 2A4 $C_{24}H_{30}N_4O_3$ Exact Mass: 422.23	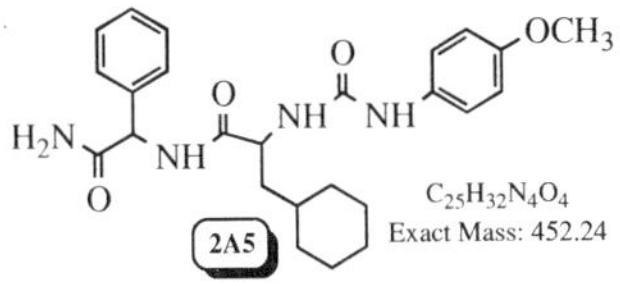 2A5 $C_{25}H_{32}N_4O_4$ Exact Mass: 452.24
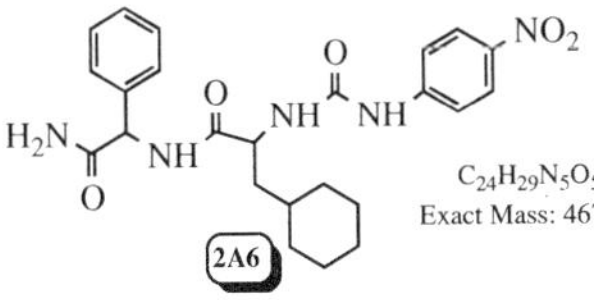 2A6 $C_{24}H_{29}N_5O_5$ Exact Mass: 467.22	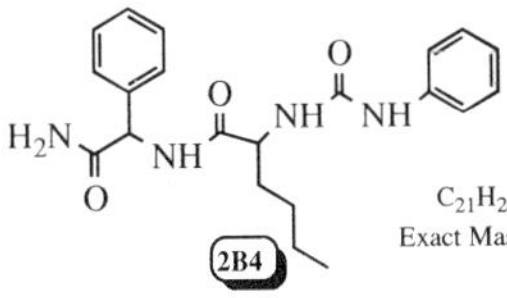 2B4 $C_{21}H_{26}N_4O_3$ Exact Mass: 382.20
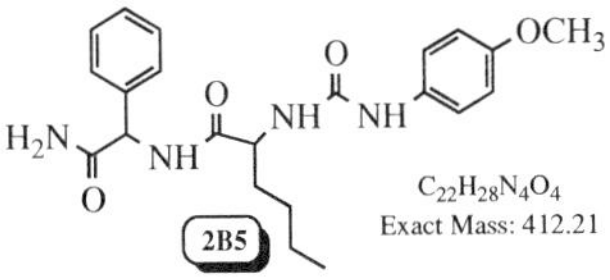 2B5 $C_{22}H_{28}N_4O_4$ Exact Mass: 412.21	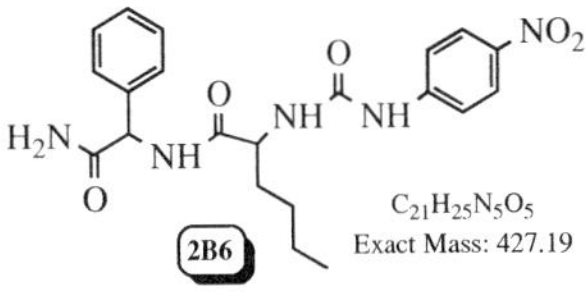 2B6 $C_{21}H_{25}N_5O_5$ Exact Mass: 427.19
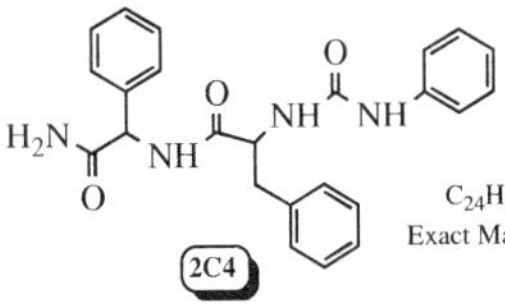 2C4 $C_{24}H_{24}N_4O_3$ Exact Mass: 416.18	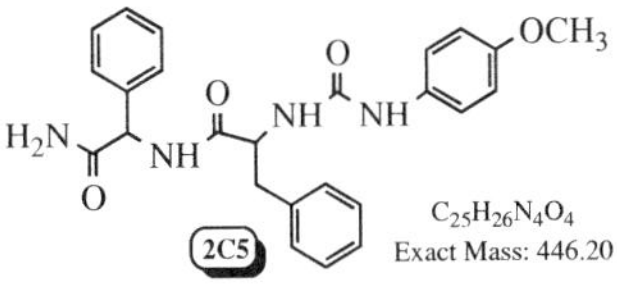 2C5 $C_{25}H_{26}N_4O_4$ Exact Mass: 446.20
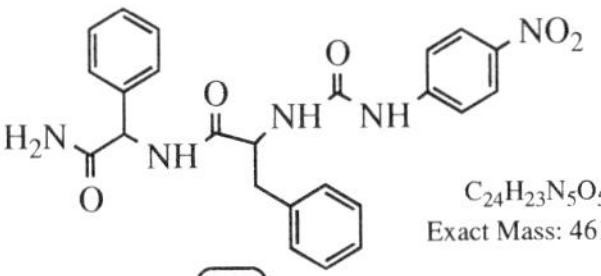 2C6 $C_{24}H_{23}N_5O_5$ Exact Mass: 461.17	

Appendix 3.3 (*Continued*)

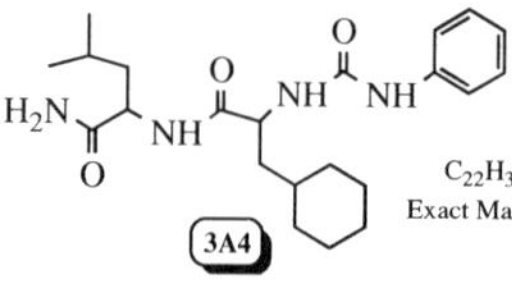

3A4
C$_{22}$H$_{34}$N$_4$O$_3$
Exact Mass: 402.26

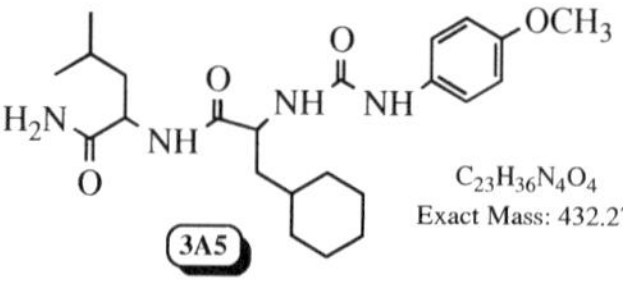

3A5
C$_{23}$H$_{36}$N$_4$O$_4$
Exact Mass: 432.27

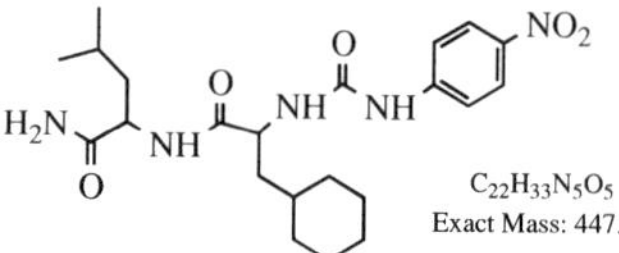

3A6
C$_{22}$H$_{33}$N$_5$O$_5$
Exact Mass: 447.25

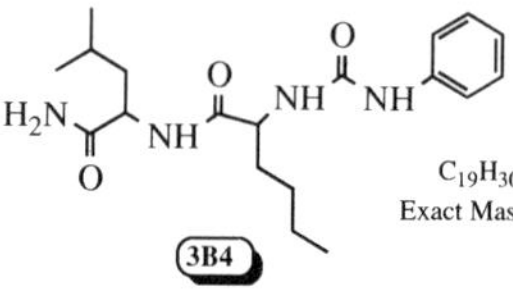

3B4
C$_{19}$H$_{30}$N$_4$O$_3$
Exact Mass: 362.23

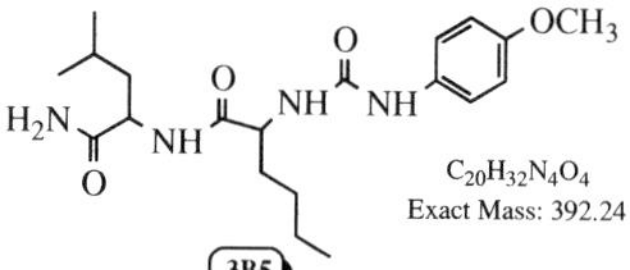

3B5
C$_{20}$H$_{32}$N$_4$O$_4$
Exact Mass: 392.24

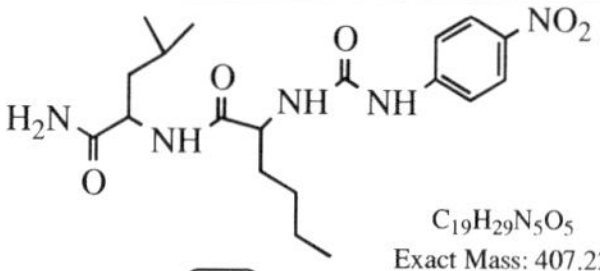

3B6
C$_{19}$H$_{29}$N$_5$O$_5$
Exact Mass: 407.22

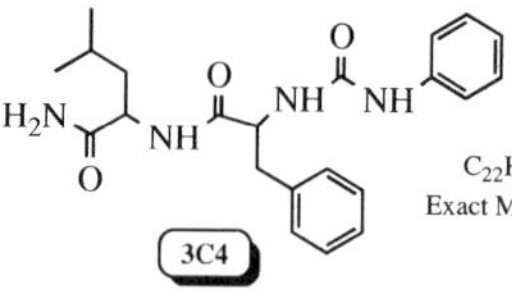

3C4
C$_{22}$H$_{28}$N$_4$O$_3$
Exact Mass: 396.22

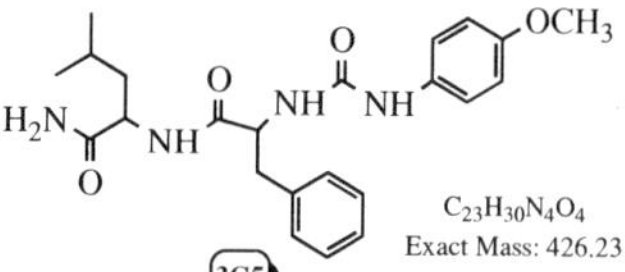

3C5
C$_{23}$H$_{30}$N$_4$O$_4$
Exact Mass: 426.23

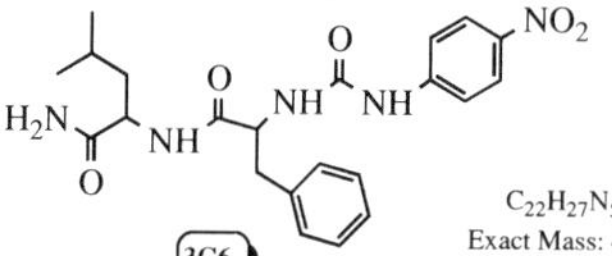

3C6
C$_{22}$H$_{27}$N$_5$O$_5$
Exact Mass: 441.20

CHAPTER FOUR

SYNTHESIS OF *p*-BENZYLOXYBENZYL CHLORIDE RESIN

Submitted by **JOHN ELLINGBOE, DEREK COLE,**
and **JOSEPH STOCK**

Wyeth-Ayerst Research, Division of Chemical Sciences,
401 North Middletown Road, Pearl River, NY, USA 10965

Checked by **KATHLEEN LIGSAY, KEVIN SHORT,**
and **TODD JONES**

Ontogen Corporation, 2325 Camino Vida Roble,
Carlsbad, CA, USA 92009

REACTION SCHEME

Wang resin   *p*-Benzyloxybenzyl chloride resin

Todd Jones current address: The R. W. Johnson Pharmaceutical Research Institute, 3210 Merryfield Row, San Diego, CA, USA 92121

PROCEDURE

Lithium chloride (Aldrich, 99%) (1.27 g, 30 mmol) was added to a suspension of Wang resin (AnaSpec Inc. Cat. # 22990, 100–200 mesh, Lot # AM5000) (10 g, 1.0 m Eq/g) in DMF (100 mL) in a 500-mL Erlenmeyer flask. 2,4,6-Collidine (Aldrich, 99%) (4.0 mL, 30 mmol) was added, followed by slow addition (over about 5 min) of methanesulfonyl chloride (Aldrich, 98%) (2.3 mL, 30 mmol; note 1). The flask was flushed with N_2, stoppered, and allowed to mix overnight on an orbital shaker (note 2). The mixture was then filtered and washed with the following solvents: 2 × 9:1 DMF:H_2O, 1 × DMF, 1 × DCM, 1 × MeOH, 2 × DMF, 2 × DCM. A wash consisted of suspending the resin in the solvent (~50 mL), stirring or swirling, then filtering. The resin was then dried *in vacuo* to give 10.1 g.

The resin was characterized by high-resolution magic angle spinning (HRMAS) NMR (Bruker 500 MHz): ^{1}H NMR ($CDCl_3$) δ 1.43 (br s), 1.84 (br s), 2.83 (s), 2.87 (s), 2.94 (m), 4.51 (s), 4.91 (br s), 5.16 (s), 5.30 (m), 6.56 (br s), 7.03 (br s), 7.98 (d); ^{13}C NMR ($CDCl_3$) δ 40.3, 46.1 (CH_2Cl), 70.0, 76.7, 76.9, 77.0, 77.2, 114.3, 115.6, 125.0, 126.2, 127.3, 128.4, 129.4, 129.8, 130.5, 133.9, 145.2, 158.9. Chlorine analysis: calculated, 3.47%; observed, 3.42%.

NOTES

1. The reaction warms slightly after the addition of methane-sulfonyl chloride. For larger scale reactions, an ice bath is used during the addition.

2. Mechanical stirring can also be used.

DISCUSSION

Polystyrene resin with a hydroxymethylphenoxy linker (Wang resin)[1] was originally developed for solid-phase peptide synthesis but has proven to be useful for solid-phase organic synthesis as well. *p*-Benzyloxybenzyl chloride resin is useful for cases in which a Wang linker is needed but when attachment to the resin can only be achieved by nucleophilic displacement of a leaving group. For example, anthranilic acid cannot be attached to Wang resin with a carbodiimide because of side reactions involving the aniline nitrogen. However, the cesium salt of anthranilic acid can be directly attached to the Wang linker via the chloro derivative, without protection of the nitrogen. This approach has been extended to other aminobenzoic acids,[2] phenols,[3] and *N*-hydroxyphthalimide (which can be converted to a hydroxylamine resin).

A synthesis of *p*-benzyloxybenzyl chloride resin using $PPh_3 \cdot Cl_2$ has been reported,[4] and $PPh_3 \cdot Br_2$ has been used to prepare a bromo Wang resin.[4, 5] Methods utilizing thionyl chloride or methanesulfonyl chloride/diisopropylethylamine have been reported more recently.[6] The combination of methanesulfonyl chloride and lithium chloride described above provides a less expensive alternative and does not produce the triphenylphosphine byproduct.

REFERENCES

1. Wang, S.-S. *J. Am. Chem. Soc.* **1973**, *95*, 1328.

2. Collini, M. D.; Ellingboe, J. W. *Tetrahedron Lett.* **1997**, *38*, 7963.

3. Chiu, C.; Tang, Z.; Ellingboe, J. W. *J. Comb. Chem.* **1999**, *1*, 73.

4. Mergler, M.; Tanner R.; Gosteli, J. *Tetrahedron Lett.* **1988**, *29*, 4005.

5. Ngu, K.; Patel, D. V. *Tetrahedron Lett.* **1997**, *38*, 973.

6. Raju, B.; Kogan, T. P. *Tetrahedron Lett.* **1997**, *38*, 4965.

CHAPTER FIVE

SOLID-PHASE MANNICH REACTIONS OF A RESIN-IMMOBILIZED ALKYNE

Submitted by SCOTT L. DAX and MARK A. YOUNGMAN

Drug Discovery, The R. W. Johnson Pharmaceutical Research Institute, Welsh and McKean Roads, Spring House, PA, USA 19477

Checked by PETR KOCIS and MATTHEW NORTH

International Lead Drug Discovery Department, Zeneca Pharmaceuticals, 1800 Concord Pike, Wilmington, DE, USA 19850-5437

LIBRARY SYNTHESIS ROUTE

a: propargyl amine (8 molar Eq.) / DMF.

b: 20 Eq. aldehyde, 10 Eq. amine, 5 Eq. 1,4-dimethylpiperazine, 1 Eq. Cu(I)Cl, dioxane, 75°C.

c: TFA / DCM (1:3).

BUILDING BLOCKS

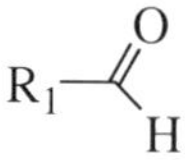

Aldehyde
component:

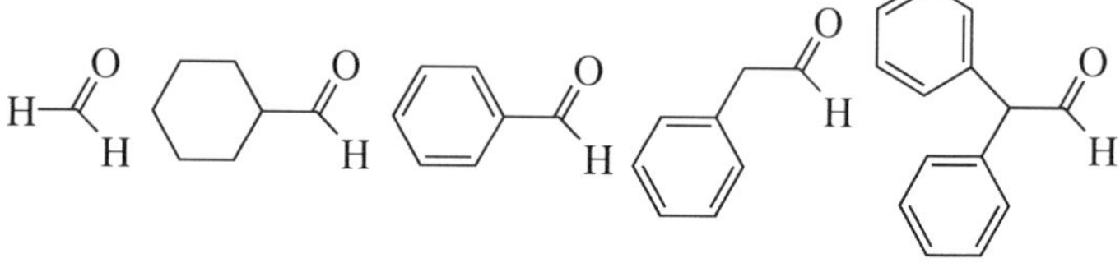

A   B   C   D   E

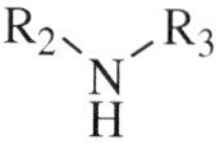

Amine
component:

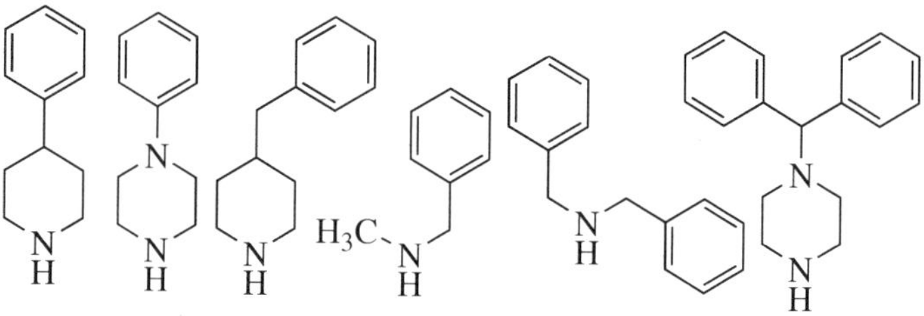

1   2   3   4   5   6

PROCEDURES

2-Cl Trityl Chloride resin (17.33 g, NovaBiochem, Lot # A20915, 200–400 mesh, 1% DVB, 1.33 mmol / g, 23.0 mmol) was placed in a 500-mL round-bottom flask. *N,N*-Dimethylformamide (200 mL) was added, which caused the resin to swell; this suspension was gently stirred by a magnetic stir bar. Propargyl amine (10 g,

180 mmol) was added, and the reaction vessel was flushed with argon, capped, and stirred gently for 20 h. The resin was removed by filtration through a sintered glass funnel and washed with DMF (3 × 100 mL) and then with methylene chloride (3 × 100 mL). The resin was dried under vacuum overnight to remove residual solvents. The 2-Cl trityl resin-bound propargyl amine prepared in this manner has a theoretical loading of 1.30 mmol/g.

The 2-Cl Trityl resin-bound propargyl amine described above was mixed with DMF-dichloroethane (3:7 v/v) with gentle stirring to provide a uniform suspension of the resin. Using a wide-bore pipette, a calculated volume of this suspension was transferred to each reaction vessel to provide 0.077 g (0.10 mmol) of the resin-bound propargyl amine. Each portion of resin was then rinsed with methylene chloride (2 × 4 mL) and air dried. Copper(I) chloride (0.010–0.015 g, 0.10–0.15 mmol; note 1) was added to each reaction vessel followed by dioxane (1 mL) and 1,4-dimethylpiperazine (0.068 mL, 0.50 mmol; note 2). The aldehyde component was added (2 mL of a 1.0 M solution or suspension in the case of formaldehyde) followed by the amine component (1 mL of a 1.0 M solution in dioxane), thus bringing the final volume of each reaction to 4 mL. The reaction vessels were capped, agitated, and heated at 75°C for approximately 6 h. After cooling the resins were filtered and washed sequentially with dioxane (1 × 4 mL), 10% piperidine in DMF (v/v) (4 × 4 mL), 5% aqueous acetic acid (1 × 4 mL), 10% piperidine in DMF (1 × 4 mL), methanol (3 × 4 mL), and methylene chloride (3 × 4 mL).

The Mannich products were cleaved from the resin into tared tubes by reaction with 4 mL of 25% trifluoroacetic acid in methylene chloride (v/v) at ambient temperature for 1 min. Each resin was filtered and rinsed with methylene chloride (2 mL). The filtrate was concentrated under a stream of nitrogen gas to a brown residue. This material was dissolved in acetonitrile (4 mL), and the product was concentrated again under a stream of nitrogen. This procedure was repeated two more times using methanol

(4 mL) to dissolve the residue (note 3). The resultant products were dried under vacuum overnight, and the tubes were weighed to obtain the final yields of the products (Tables 5.1 and 5.2). The products were typically obtained as brown glassy solids (note 4). A portion of the solid was removed and dissolved in methanol for HPLC and MS analysis. The remainder of the product was dissolved in d_4-methanol for NMR analysis.

TABLE 5.1.   Percent Yields[a]

Component	A	B	C	D	E
1	57/46	51/54	34/53	63/55	36/62
2	57/57	68/65	62/65	42/66	24/74
3	61/47	59/55	58/55	66/56	37/64
4	51/42	44/50	15/49	42/51	28/58
5	60/49	65/57	41/57	25/58	21/66
6	75/66	87/74	105/74	58/75	27/83

[a] Isolated weight (mg)/theoretical weight (mg).

TABLE 5.2.   Purity[a]

Component	A	B	C	D	E
1	>95%	>95%	95%	>95%	>95%
2	>95%	>95%	>95%	>95%	45%
3	>95%	>95%	>95%	>95%	>95%
4	>95%	>95%	>95%	>95%	71%
5	>95%	>95%	80%	47%	17%
6	>95%	>95%	69%	>95%	53%

[a] Determined by reverse-phase HPLC (acetonitrile–water gradient containing 0.1% TFA; 220 nM).

NOTES

1. Copper (I) chloride was ground to a fine powder using a mortar and pestle before use.

2. We have observed that 1,4-dimethylpiperazine is an innocuous additive that improves both the yield and the crude purity of some Mannich products. Accordingly, dimethylpiperazine was used in this array to provide uniform reaction conditions, although it is not needed for the formation and isolation of many Mannich adducts in this library.

3. Trace amounts of unreacted propargyl amine were observed to be the lone impurity in some reactions.

4. Final products were isolated as solid glasses and typically contained minor amounts of residual methanol and water ($\sim$5 to $\sim$25%).

DISCUSSION

Multicomponent reaction systems are highly valued in solid-phase organic synthesis because several elements of diversity can be introduced in a single transformation.[1] The Mannich reaction is a classic example of a three-component system in which an "active hydrogen" component, such as a terminal alkyne, undergoes condensation with the putative imine species formed from the condensation of an amine with an aldehyde.[2] The resultant Mannich adducts contain at least three potential sites for diversification; specifically, each individual component—the amine, aldehyde, and alkyne—can be varied in structure and thus provide an element of diversity.

We describe here Mannich reactions of a resin-immobilized alkyne and demonstrate the versatility of this methodology.[3] Aryl-, alkyl-, aralkyl-aldehydes, and formaldehyde are suitable aldehyde components; both cyclic and acyclic secondary amines are amenable to this chemistry. A $1 \times 5 \times 6$ library is reported; formation of the Mannich adducts generally proceeded in good yield and the purity of the crude products was typically excellent.

REFERENCES

1. Dax, S. L.; McNally, J. J.; Youngman, M. A. *Curr. Med. Chem.* **1999**, *6*, 251.

2. Tramontini, M.; Angiolini, L. *Mannich Bases: Chemistry and Uses*, CRC Press: Boca Raton, Fla., 1994.

3. Youngman, M. A.; Dax, S. L. *Tetrahedron Lett.* **1997**, *38*, 6347.

Appendix 5.1

Experimental Supplement

Compound A1. ^{1}H NMR (CD$_3$OD) 7.37–7.18 (m, 5H), 4.21 (s, 2H), 3.97 (s, 2H), 3.74 (d, 2H), 3.39–3.21 (m, 2H), 2.99–2.81 (m, 1H), 2.20–1.97 (m, 4H); ES-MS m/z 229(MH$^+$); C$_{15}$H$_{20}$N$_2 \cdot$ 2TFA (456.38).

Compound A2. ^{1}H NMR (CD$_3$OD) 7.37–7.22 (m, 2H), 7.03 (d, 2H), 6.93 (t, 1H), 4.26 (s, 1H), 3.97 (s, 2H), 3.69–3.23 (br m, 8H); ES-MS m/z 230(MH$^+$); C$_{14}$H$_{19}$N$_3 \cdot$ 3TFA (571.39).

Compound A3. ^{1}H NMR (CD$_3$OD) 7.29 (m, 2H), 7.19 (m, 3H), 4.12 (s, 2H), 3.92 (s, 2H), 3.60 (br d, 2H), 3.18–2.93 (m, 2H), 2.63 (d, 2H), 2.00–1.77 (m, 3H), 1.67-1.42 (m, 2H); ES-MS m/z 243(MH$^+$); C$_{16}$H$_{22}$N$_2 \cdot$ 2TFA (470.41).

Compound A4. ^{1}H NMR (CD$_3$OD) 7.60–7.44 (m, 5H), 4.42 (s, 2H), 4.11 (s, 2H), 4.00 (s, 2H) 2.94 (s, 3H); ES-MS m/z 189(MH$^+$); C$_{12}$H$_{16}$N$_2 \cdot$ 2TFA (416.31).

Compound A5. ^{1}H NMR (CD$_3$OD) 7.57 (m, 4H), 7.48 (m, 6H), 4.47 (s, 4H), 4.07 (s, 2H), 3.81 (s, 2H); ES-MS m/z 265(MH$^+$); C$_{18}$H$_{20}$N$_2 \cdot$ 2TFA (492.41).

Compound A6. ^{1}H NMR (CD$_3$OD) 7.57 (d, 4H), 7.42–7.21 (m, 6H), 4.77 (s, 1H), 3.98 (d, 2H), 3.90 (s, 2H), 3.31 (br d, 4H), 2.93 (br, 4H); ES-MS m/z 320(MH$^+$); C$_{21}$H$_{25}$N$_3 \cdot$ 3TFA (661.51).

Compound B1. ^{1}H NMR (CD$_3$OD) 7.39-7.77 (m, 5H), 4.20 (br d, 1H), 4.03 (s, 2H), 3.86 (br d, 1H), 3.65 (br d, 1H), 3.52–3.49 (m, 1H), 3.48-3.22 (m, 1H), 3.02-2.82 (m, 1H), 2.27–1.92 (m, 6H), 1.91–1.66 (m, 4H), 1.49–1.13 (m, 5H); ES-MS m/z 311(MH$^+$); C$_{21}$H$_{30}$N$_2 \cdot$ 2TFA (538.53).

Compound B2. ^{1}H NMR (CD$_3$OD) 7.32 (t, 2H), 7.08 (d, 2H), 6.98 (t, 1H), 4.18 (d, 1H), 3.99 (s, 2H), 3.76–3.34 (br m, 8H), 2.12–1.92 (m, 2H), 1.91–1.64 (m, 4H), 1.48–1.14 (m, 5H); ES-MS m/z 312(MH$^+$); C$_{20}$H$_{29}$N$_3 \cdot$ 3TFA (653.53).

Compound B3. ^{1}H NMR (CD$_3$OD) 7.29 (t, 2H), 7.19 (m, 3H), 4.10 (br d, 1H), 3.97 (s, 2H), 3.72 (br d, 1H), 3.50 (br d, 1H), 3.23 (br t, 1H), 3.07 (br t, 1H), 2.60 (d, 2H), 2.05–1.48 (m, 11H), 1.47–1.10 (m, 5H); ES-MS m/z 325(MH$^+$); C$_{22}$H$_{32}$N$_2 \cdot$ 2TFA (552.55).

Compound B4. ^{1}H NMR (CD$_3$OD) 7.67–7.41 (m, 5H), 4.60–4.43 (m, 1H), 4.42–4.25 (m, 1H), 4.09–3.88 (m, 3H), 2.87 (s, 3H), 2.09–1.91 (m, 2H), 1.90–1.46 (m, 4H), 1.45–0.96 (m, 5H); ES-MS m/z 271(MH$^+$); C$_{18}$H$_{26}$N$_2 \cdot$ 2TFA (498.46).

Compound B5. ^{1}H NMR (CD$_3$OD) 7.55–7.22 (m, 10H), 4.13–3.92 (m, 3H), 3.82–3.63 (d, 2H), 3.34–3.22 (d, 2H), 2.19–1.91 (br dd, 2H), 1.83–1.49 (m, 4H), 1.37–0.98 (m, 3H), 0.97–0.64 (m, 2H); ES-MS m/z 347(MH$^+$); C$_{24}$H$_{30}$N$_2 \cdot$ 2TFA (574.56).

Compound B6. ^{1}H NMR (CD$_3$OD) 7.77–7.55 (m, 4H), 7.49–7.21 (m, 6H), 5.23 (s, 1H), 3.87 (s, 2H), 3.42–3.26 (m, 1H), 3.25–2.98 (m, 6H), 2.97–2.73 (m, 2H), 2.10–1.85 (m, 2H), 1.84–1.47 (m, 4H), 1.49–1.13 (m, 3H), 1.12–0.85 (m, 2H); ES-MS m/z 402(MH$^+$); C$_{27}$H$_{35}$N$_3$ · 3TFA (743.66).

Compound C1. ^{1}H NMR (CD$_3$OD) 7.73 (m, 2H), 7.55 (m, 3H), 7.37–7.17 (m, 5H), 5.68 (s, 1H), 4.07 (s, 2H), 3.71 (br d, 1H), 3.60 (br d 1H), 3.39–3.19 (m, 2H), 2.95–2.80 (m, 1H), 2.18–1.92 (m, 4H); ES-MS m/z 305(MH$^+$); C$_{21}$H$_{24}$N$_2$ · 2TFA (532.48).

Compound C2. ^{1}H NMR (CD$_3$OD) 7.72 (m, 2H), 7.53 (m, 3H), 7.31 (t, 2H), 7.08 (d, 2H), 6.99 (t, 1H), 5.63 (s, 1H), 4.07 (s, 2H), 3.65–3.33 (br m, 8H); ES-MS m/z 306(MH$^+$); C$_{20}$H$_{23}$N$_3$ · 3TFA (647.49).

Compound C3. ^{1}H NMR (CD$_3$OD) 7.67 (m, 2H), 7.53 (m, 3H), 7.27 (m, 2H), 7.17 (m, 3H), 5.60 (s, 1H), 4.02 (s, 2H), 3.60 (br s, 1H), 3.48 (br d, 1H), 3.20–2.98 (m, 2H), 2.59 (d, 2H), 2.02–1.74 (m, 3H), 1.67–1.39 (m, 2H); ES-MS m/z 319(MH$^+$); C$_{22}$H$_{26}$N$_2$ · 2TFA (546.51).

Compound C4. ES-MS m/z 265(MH$^+$); C$_{18}$H$_{20}$N$_2$·2TFA (492.41).

Compound C5. ES-MS m/z 341(MH$^+$); C$_{24}$H$_{24}$N$_2$ · 2TFA (568.51).

Compound C6. ES-MS m/z 396(MH$^+$); C$_{27}$H$_{29}$N$_3$ · 3TFA (737.61).

Compound D1. ^{1}H NMR (CD$_3$OD) 7.47–7.14 (m, 10H), 4.62 (br d, 1H), 3.93 (s, 2H), 3.89–3.78 (m, 1H), 3.77–3.64 (m, 1H), 3.63–3.28 (m, 3H), 3.27–3.05 (m, 1H), 3.04–2.84 (m, 1H), 2.28–2.03 (m, 4H); ES-MS m/z 319(MH$^+$); C$_{22}$H$_{26}$N$_2$ · 2TFA (546.51).

Compound D2. ^{1}H NMR (CD$_3$OD) 7.43–7.18 (m, 7H), 7.07 (d, 2H), 6.97 (t, 1H), 4.62 (br d, 1H), 3.81 (s, 2H), 3.74–3.43 (m, 8H), 3.23–3.04 (m, 2H); ES-MS *m/z* 320(MH$^+$); C$_{21}$H$_{25}$N$_3$ · 3TFA (661.51).

Compound D3. ^{1}H NMR (CD$_3$OD) 7.40–7.24 (m, 7H), 7.20 (d, 3H), 4.53 (br d, 1H), 3.88 (s, 2H), 3.71 (br d, 1H), 3.58 (br d, 1H), 3.39–3.17 (m, 3H), 3.11 (t, 1H), 2.64 (d, 2H), 2.08–1.84 (m, 3H), 1.74–1.52 (m, 2H); ES-MS *m/z* 333(MH$^+$); C$_{23}$H$_{28}$N$_2$ · 2TFA (560.53).

Compound D4. ^{1}H NMR (CD$_3$OD) 7.58 (m, 2H), 7.50 (s, 3H), 7.32 (m, 5H), 4.61–4.38 (m, 3H), 3.98 (br s, 2H), 3.27–3.07 (m, 2H), 2.91 (s, 3H); ES-MS *m/z* 279(MH$^+$); C$_{19}$H$_{22}$N$_2$ · 2TFA (506.44).

Compound D5. ES-MS *m/z* 355(MH$^+$); C$_{25}$H$_{26}$N$_2$ · 2TFA (582.54).

Compound D6. ES-MS *m/z* 410(MH$^+$); C$_{28}$H$_{31}$N$_3$ · 3TFA (751.64).

Compound E1. ES-MS *m/z* 395(MH$^+$); C$_{28}$H$_{30}$N$_2$ · 2TFA (622.60).

Compound E2. ES-MS *m/z* 396(MH$^+$); C$_{27}$H$_{29}$N$_3$ · 3TFA (737.61).

Compound E3. ES-MS *m/z* 409(MH$^+$); C$_{29}$H$_{32}$N$_2$ · 2TFA (636.63).

Compound E4. ES-MS *m/z* 355(MH$^+$); C$_{25}$H$_{26}$N$_2$ · 2TFA (582.54).

Compound E5. ES-MS *m/z* 431(MH$^+$); C$_{31}$H$_{30}$N$_2$ · 2TFA (658.64).

Compound E6. ES-MS *m/z* 486(MH$^+$); C$_{34}$H$_{35}$N$_3$ · 3TFA (827.74).

CHAPTER SIX

SOLID-PHASE SYNTHESIS OF DI-β-PEPTOIDS FROM ACRYLATE RESIN: *N*-ACETYL-*N*-BENZYL-β-ALANINYL-*N*-BENZYL-β-ALANINE

Submitted by **BRUCE C. HAMPER** and
ALLEN S. KESSELRING

Searle, Parallel Medicinal Chemistry, Monsanto Company,
800 North Lindbergh Boulevard, St. Louis, MO, USA 63167

Checked by **MARSHALL H. PARKER** and **JAMES A. TURNER**

Dow AgroSciences LLC, 9330 Zionville Road,
Indianapolis, IN, USA 46268-1054

REACTION SCHEME

A. (resin)–OH + (acryloyl chloride) $\xrightarrow{\text{Et}_3\text{N, CH}_2\text{Cl}_2}$ (resin acrylate) **1**

B. **1** + (benzylamine) $\xrightarrow[\text{25°C}]{\text{DMSO}}$ (resin β-alanine benzyl) **2**

C. **2** + [acryloyl chloride] $\xrightarrow{\text{Et}_3\text{N, CH}_2\text{Cl}_2}$ [resin-bound N-benzyl acrylamide ester] **3**

D. **3** + [benzylamine] $\xrightarrow[\text{55°C}]{\text{DMSO}}$ [resin-bound product] **4**

E. **4** + [acetic anhydride] $\xrightarrow{\text{Et}_3\text{N, DMF}}$ [resin-bound acetylated product] **5**

F. **5** $\xrightarrow{\text{TFA}}$ [cleaved product] **6**

PROCEDURE

Acrylate Resin (1)

To 11.0 g (12.3 mmol) of Wang resin (Note 1) in an oven-dried, solid-phase reaction flask (Note 2) equipped with an overhead stirrer and a nitrogen line attached to a bubbler was added 100 mL dichloromethane. The resultant slurry was allowed to stir for

10 min at room temperature and subsequently treated with 4.3 mL (30.8 mmol) of triethylamine followed by dropwise addition of a solution of 2.0 mL (24.6 mmol) acryloyl chloride in 4 mL of dichloromethane (Note 3). After stirring for 2 h at room temperature, the nitrogen line was removed and the flask contents were filtered in the vessel by attaching a vacuum line equipped with a trap to the sidearm and opening the Teflon stopcock. The resin was washed with an additional 50 mL dichloromethane, allowed to stir for 2 min, the solvent was removed by suction. To ensure completion of the reaction, the resin was subjected to a second treatment with a solution containing 100 mL dichloromethane, 4.3 mL triethylamine (30.8 mmol), and 2.0 mL acryloyl chloride (24.4 mmol) and stirred for 2 h. The resin was filtered and washed three times with 50 mL *each* of the following solvents: dichloromethane, methanol, N,N-dimethylacetamide, methanol, and dichloromethane. After completion of the washing steps, the vacuum line was removed from the sidearm, and a nitrogen line was attached to allow for a positive flow of nitrogen to induce drying of the resin. After 24 h, the nitrogen line was removed and a 70 mg sample removed for determination of loading by direct cleavage ^{1}H NMR (Note 4). The acrylate resin (**1**) was obtained as a light yellow solid: FTIR (KBr) 1725 cm^{-1} (C=O). Loading was determined by direct cleavage ^{1}H NMR: 0.98 mEq/g (theoretical, 1.06 mEq/g; yield, 93%; Note 5).

N-Benzyl-β-Alanine-Wang Resin (2)

To product **1** (11.58 g [calculated], 11.39 mmol) in the reaction flask from the above procedure was added 50 mL methyl sulfoxide and 7.5 mL (68.3 mmol) benzylamine (Note 6) and the slurry was allowed to stir for 24 h at room temperature. The resin was filtered, retreated with 50 mL methyl sulfoxide and 7.5 mL (68.3 mmol) benzylamine, and stirred for another 24 h at room temperature. The reagents were removed by suction filtration in the vessel, the resin was washed three times *each* with 50 mL portions of *N,N*-dimethylacetamide, methanol, and dichloromethane, and the washed resin was dried by applying a stream of nitrogen to the vessel overnight. *N*-Benzyl-β-alanine (**2**) was obtained as a yellow resin: FTIR (KBr) 1733 cm^{-1} (C=O); direct cleavage (68.8 mg **2** with 1.00 mL standard cleavage solution) ^{1}H NMR (CDCl$_3$/TFA) δ 2.96 (t, 2H, 6.0 Hz), 3.42 (m, 2H), 4.35 (t, 2H, 5.5 Hz), 7.37 (m, 2H), 7.48 (m, 3H), 7.75 (broad s, 2H), integral regions: HMDS 0.42 (10.0 counts, 18 H), 2.96 (6.20 counts, 2H); calculated, loading, 0.755 mEq/g (theoretical, 0.889 mEq/g; yield, 84.9%; Note 7).

N-Acryloyl-*N*-Benzyl-β-Alanine-Wang Resin (3)

To product **2** (12.80 g [calculated], 9.67 mmol) in the reaction flask from last procedure was added 100 mL of dichloromethane and 3.4 mL (24.4 mmol) triethylamine. The slurry was stirred at room temperature and treated dropwise with 1.57 mL (19.3 mmol) acryloyl chloride. After the addition was complete, the mixture was allowed to stir at room temperature for 2 h. The resin was filtered by suction in the reaction flask; washed with 50 mL dichloromethane; and retreated with 50 mL dichloromethane, 3.4 mL (24.4 mmol) triethylamine, and 1.57 mL (19.3 mmol) acryloyl chloride. This second treatment was allowed to stir for 2 h and was then filtered and washed three times with 50 mL *each* of the following solvents: dichloromethane, methanol, *N,N*-dimethylacetamide, methanol, and dichloromethane. The reaction vessel was flushed with nitrogen to allow drying of the resin overnight to afford *N*-acryloyl-*N*-benzyl-β-alanine resin (**3**) as a light, yellow solid: FTIR (KBr) 1733 (C=O, ester) and 1652 cm^{-1} (C=O, amide); direct cleavage (99.4 mg **3** with 1.00 mL standard cleavage solution) ^{1}H NMR (CDCl$_3$/TFA) δ 2.80 (m, 2H), 3.81 (m, 2H), 4.77 (m, 2H), 5.99 (m, 1H), 6.43 (m, 1H), 6.62 (m, 1H), 7.18 ≃ 7.42 (m, 5H), integral regions: HMDS 0.42 (10.0 counts, 18 H), 2.80 (8.03 counts, 2H), 3.81 (7.69 counts, 2H); calculated loading, 0.661 mEq/g (theoretical, 0.695 mEq/g; yield, 95%; Note 7).

N-Benzyl-*β*-Alaninyl-*N*-Benzyl-*β*-Alanine Wang Resin (4)

To product **3** (13.33 g [calc], 8.93 mmol) in the reaction flask from procedure C, was added 50 mL methyl sulfoxide and 11.7 mL (107.2 mmol) of benzylamine (Note 6). A heating mantle was added under the reaction flask and the stirred slurry heated to 55°C for 24 h. After removing the heating mantle and allowing the mixture to cool to rt, the resin was filtered and retreated with 50 mL of methyl sulfoxide and 11.7 mL (107.2 mmol) of benzylamine. The slurry is stirred for another 24 h at 55°C, cooled to 20°C with the aid of a water bath, filtered in the vessel and the resultant resin thoroughly washed three times with 50 mL portions of each of the following solvents: *N,N*-dimethylacetamide, methanol, and dichloromethane. The product was dried overnight under a stream of nitrogen to afford *N*-benzyl-*β*-alaninyl-*N*-benzyl-*β*-alanine resin (**4**): FTIR (KBr) 1733 (C=O, ester) and 1648 cm^{-1} (C=O, amide); direct cleavage (103.5 mg **4** with 1.00 mL standard cleavage solution) ^{1}H NMR (CDCl$_3$/TFA) mixture of two conformers: δ 2.70 (m, 2H), 2.98 (m, 2H), 3.39 (m, 2H), 3.75 (m, 2H), 4.32 (m, 2H), 4.61 (m, 2H), 7.07–7.48 (m, 10H), 7.77 (broad s, 2H), integral regions: HMDS 0.42 (10.0 counts, 18H), 2.70 (3.67 counts, 2H) 4.32 (4.03 counts, 2H); calculated loading, 0.623 mEq/g (theor. 0.625 mEq/g; yield, 99.7%; Note 7).

N-Acetyl-*N*-Benzyl-*β*-Alaninyl-*N*-Benzyl-*β*-Alanine Wang Resin (5)

To a slurry of product **4** (13.07 g [calculated], 8.14 mmol) in 100 mL DMF in the reaction flask from the last procedure was added 5.85 mL (62.1 mmol) acetic anhydride (Note 8) and 8.66 mL (62.1 mmol) triethylamine. The slurry was stirred for 3 h at room temperature. The resin was filtered and washed three times each with 50 mL portions of *each* of the following solvents: *N,N*-dimethylacetamide, methanol, and dichloromethane. After drying the resin by allowing nitrogen to flow through the reaction vessel overnight, *N*-acetyl-*N*-benzyl-β-alaninyl-*N*-benzyl-β-alanine Wang resin (**5**) was obtained as a yellow resin: direct cleavage (124.6 mg **5** with 1.00 mL standard cleavage solution) ^{1}H NMR (CDCl$_3$/TFA) mixture of conformers: δ 2.40 (m, 3H), 2.72 (m, 4H), 3.77 (m, 4H), 4.68 (m, 4H), 7.10–7.40 (m, 10H), integral regions: HMDS 0.42 (10.0 counts, 18 H), 2.40 (12.2 counts, 3H) 2.72 (3.94 counts, 4H); calculated loading, 0.536 mEq/g (theoretical, 0.607 mEq/g; yield, 88.3% Note 7).

N-Acetyl-*N*-Benzyl-β-Alaninyl-*N*-Benzyl-β-Alanine (6)

Product **5** (13.3 g [calculated], 7.12 mmol) was treated with 100 mL trifluoroacetic acid:water (95:5) and allowed to stir for 45 min at room temperature. The resin was transferred to a 500-mL round bottom flask, filtered through a course sintered glass frit, washed three times with 50 mL portions of methylene chloride and the combined filtrates concentrated in vacuo to afford 4.18 g of a crude yellow oil. The highly viscous oil retains solvent, which is difficult to remove without extensive drying in a vacuum oven; however, the purity is >85% as determined by ^{1}H NMR and LCMS analysis. The crude oil was purified by preparative scale reverse-phase chromatography (C18 column, 2"×11", 70% acetonitrile:30% H$_2$O/0.1% TFA), and a heart cut of the major peak collected. This fraction was concentrated in vacuo and the product dried in a vacuum oven (50°C, 1 torr) overnight to afford 1.39–2.06 g (51.0–75.6%; overall yield for six steps is 30.6–43.6%) of *N*-acetyl-*N*-benzyl-β-alaninyl-*N*-benzyl-β-alanine **6** as a highly viscous yellow oil-glass (Note 9): ^{1}H NMR (DMSO-d_6, 300 MHz, 122°C) 2.04 (broad s, 3H), 2.44 (t, 2H, 7.2 Hz), 2.62 (broad s, 2H), 3.53 (m, 4H), 4.73 (s, 4H), 7.15–7.32 (m, 10H); MS (ESI+) 383 (M+1, 100), 384 (M+2, 22), 405 (M+Na, 42); HRMS (ESI+) *m/z* calcd for (C$_{22}$H$_{27}$N$_2$O$_4$) 383.1971, found 383.1961.

DIVERSITY REAGENTS

$$\text{HO} \overset{O}{-\!\!\!\diagup\!\!\!-} \overset{R_1}{\underset{}{N}} \overset{O}{-\!\!\!\diagup\!\!\!-} \overset{R_2}{\underset{}{N}} \overset{O}{-\!\!\!\diagup\!\!\!-} R_3$$

7

Diversity reagents for the synthesis of N-capped di-β-peptoids **7** can be introduced by the Michael addition of amines (steps B and D) for R_1 and R_2 and use of different capping groups (step F) for R_3. The availability of amines suitable for addition to the acrylate or acrylamide resins **1** and **3** allows for the synthesis of a wide variety of di-β-peptoids. Typically, addition to the acrylamide resin **3** requires higher temperatures or longer reaction times than addition to acrylate **1**. For investigation of the Michael addition, we chose 12 amines **8** for addition to acrylamide resin **3** using an Argonaut Nautilus synthesizer to carry out the parallel synthesis. This reactor allows automated control of temperature, addition of reactants, and washing of the resins. Controlled cooling of the resins after reaction before the washing step proved critical for obtaining high yields of the di-β-peptoids **9** (Note 10).

The amine diversity reagents **8a–l** were investigated in parallel by adding 100 mg N-acryloyl-N-benzyl-β-alanine Wang resin **3** (loading $= 0.670\,\text{mEq}/\text{g}$) to each of 12-8 mL Nautilus reaction vessels. Each glass vessel is equipped with two Teflon filter frits attached to an inlet and outlet, allowing flow through treatment with reagents and solvents. The vessels **a–l** were treated with 2 mL 2M solutions of the amines **8a–l** in DMSO. A neutral 2M solution of β-alanine ethyl ester was prepared by adding an excess of $NaHCO_3$ to the 2 M solution of the hydrochloride salt in DMSO. The vessels were heated to 50°C and agitated with a rocking motion in the Nautilus reaction module. After 24 h, the vessels were emptied by filtration in the reaction module and retreated with 2 M solutions of the appropriate amine. Following the 24 h second treatment, the vessels were cooled to 20°C using chilled N_2 gas and subse-

TABLE 6.1.   Preparation of Di-β-PEPTOIDS 9 FROM ACRYLAMIDE RESIN 3 AND AMINES 8

Entry	R-NH2	Yield (%)[a]	Conversion (%)[a]	HPLC (min., area %)[b]
a	benzyl-NH$_2$	85.2	>95	2.09 (93%)
b	phenethyl-NH$_2$	72.1	>95	2.42 (>95%)
c	*p*-methoxybenzyl-NH$_2$	82.7	>95	2.20 (>95%)
d	allyl-NH$_2$	74.4	>95	1.10 (92%)
e	iso-butyl-NH$_2$	92.5	>95	1.67 (91%)
f	sec-butyl-NH$_2$	77.9	>95	1.50 (94%)
g	iso-propyl-NH$_2$	81.7	>95	1.10 (91%)
h	naphthalenemethyl-NH$_2$	91.8	>95	2.59 (86%)
i	cyclopropyl-NH$_2$	73.2	90	1.03 (86%) (6% SM)
j	EtOOCCH$_2$CH$_2$-NH$_2$	72.8	83	0.98 (85%)
k	*n*-dodecyl-NH$_2$	71.2	>95	4.82 (>95%)
l	phenyl-NH$_2$	0	0	0

[a] Yield and conversion were determined by direct cleavage ^{1}H NMR (Note 13). The yield represents the percent µmoles of product compared to theoretical. Conversions were determined by comparison of the acrylamide **3** and product **9** resonances.
[b] HPLC retention times and area percent of major peak (Note 14). The acrylamide product from resin **3** has a retention time of 1.75 min.

quently washed three times each with 2 mL portions of dimethylacetamide, MeOH and CH_2Cl_2. The resin in vessel **k** was washed three times *each* with 2 mL portions of 10% aqueous acetic acid, water, dimethylacetamide, MeOH, and CH_2Cl_2 (Note 11). The resins were dried by applying a stream of N_2 for 1 h prior to direct cleavage ^{1}H NMR determination of loading and conversion (Note 12).

The reaction vessels were removed from the reaction module, placed in a shaker rack and treated with 1.00 mL of 9.3 mM HMDS in TFA/CDCl$_3$ (1:1). After shaking for 1 h, the contents of the vessels were transferred to 15 mL polypropylene vessels equipped with a filter frit, and the filtrate was collected in 4 mL analytical vials. The cleaved resins were washed three times with 0.2 mL portions of $CDCl_3$, the combined filtrates were collected and transferred to NMR tubes. A small portion of the sample was placed in an analytical vial and diluted with acetonitrile for HPLC analysis (Table 6.1).

NOTES

1. Wang resin was acquired from Chem-Impex. (1% DVB cross-linked, *p*-benyloxybenzyl alcohol resin. Grain size 100–200 mesh. Cat. #01927. Lot #N12270. Subs 1.12 mEq/g). The checkers used 10.30 g of Wang resin with a loading of 1.20 mmol/g (12.36 mmol) obtained from Midwest Biotech. Resins were dried in a vacuum dessicator before use.

2. A custom solid-phase reaction flask (250 mL) was used for preparation of the resins (Fig. 6.1), which allows for convenient washing of the resin between steps, gentle agitation with an overhead paddle stirrer, inert atmosphere, and the ability to place the vessel in heating or cooling baths. Typical resin washing steps are carried out by attaching a vacuum line equipped with a trap to the sidearm and opening the stopcock for filtration of the resin. After closing the

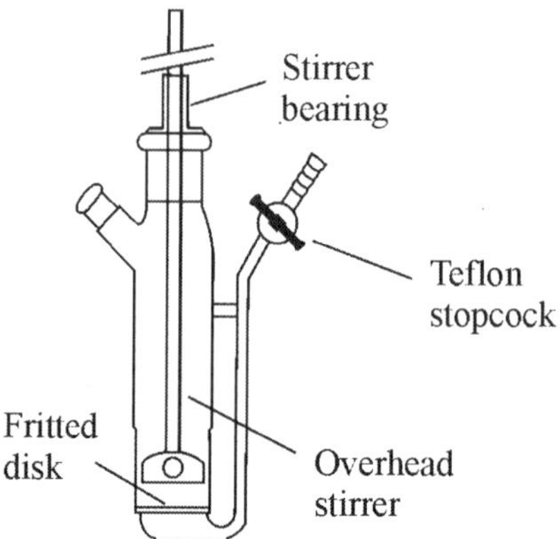

Figure 6.1. Solid-phase reaction flask.

stopcock, the wash solvent is introduced and slurried with the resin for a few minutes before filtration and addition of the next wash solvent. This design has a major advantage over the standard solid-phase peptide synthesis vessels, because the flask can be placed directly in a heating or cooling bath. A similar vessel of smaller size (15 mL) is available from Aldrich (Cat. # Z28,330-4), although a Teflon or glass stopcock is preferable to the O-ring needle valve of the commercial vessel. The checkers employed a commercial solid-phase peptide synthesis vessel (Aldrich Cat. # Z16, 229-9) which consists of a 1 L flask equipped with four S/T 24/40 joints at the top and a course sintered glass frit and stopcock at the bottom.

3. Triethylamine was obtained from Fisher Scientific Company and used without further purification. Acryloyl chloride was purchased from Aldrich Chemical Company and used without further purification.

4. Loadings of substrates on resins were determined by cleavage of the resin samples with a known quantity of hexamethyl-disiloxane (HMDS) in 50:50 TFA/CDCl$_3$ and comparison of the ^{1}H NMR integrals of the HMDS standard and the cleaved

product. A standard solution of 100 mL 9.306 mM HMDS in TFA:CDCl$_3$ (1:1) was prepared and used for all determinations of polymeric loadings. Measurement of the ^{1}H NMR integrals of the HMDS peak (0.421 ppm relative to TMS) and the product allowed direct determination of the molar concentration of cleaved product. For direct cleavage ^{1}H NMR measurement of acrylate resin **1**, a sample was dried in vacuo overnight. To 70.3 mg of **1** (dried to constant weight) was added 1.00 mL of 9.306 mM HMDS in TFA:CDCl3 (1:1) and the mixture shaken for 30 min at room temperature. The flitrate was collected using a disposable 15 mL polypropylene vessel equipped with a frit (Alltech, Cat. # 210315 and # 211412) and the resin washed three times with 0.2-mL portions of CDCl$_3$. The combined filtrates were transferred to an NMR tube for measurement of loading: ^{1}H NMR (CDCl$_3$/ TFA) δ 6.14 (m, 2H), 6.64 (dd, 1H, 16 Hz, 2 Hz); integral regions: HMDS 0.42 (18H, 13.3 counts), 6.14 (2H, 10.49 counts), 6.64 (1H, 5.49 counts). The loading of the resin was calculated from the relative integral regions as follows:

acrylic acid (µmol) = (µmol HMDS) (counts/H of acrylic acid)/(counts/H of HMDS)

= (9.306 µmol) (5.49 counts/H)/(13.3 counts/18H) = 69.1 µmol

Loading of **1** (mmol/g) = (µmol acrylic acid) / (weight of cleaved resin)

= 69.1 µmol/70.3 mg = 0.983 mmol/g

5. Theoretical loadings were determined by assuming complete conversion of the substrate attached to the resin and taking into account the change in weight of the resin. For acrylate resin **1** the theoretical loading was calculated as follows:

Theoretical loading of **1** (mEq./g) = (mmol starting resin)/(total weight of product resin)

In this case, 1 g starting resin contains 1.12 mmol Wang linker based on the reported loading from the manufacturer. Assuming complete conversion of all sites, the Wang linker—OH group would be completely replaced by the acrylate—$O(CO)CH=CH_2$ fragment. The total weight of the resin would correspond to the addition of 1.12 mmol of the difference of these two molecular fragments (C_3H_2O).

Total weight of product resin $= 1.000\,g + [(1.12 \times 10^{-3}\,mol)\,(54.049\,g/mol)]$

Theoretical loading of **1** $= (1.12\,mmol)/(1.061\,g) = 1.056\,mmol/g$

The checkers obtained an NMR calculated loading of 1.07 mmol/g (theoretical, 1.127 mEq./g; yield, 95%).

6. The checkers employed 75 mL DMSO rather than 50 mL to facilitate slurry agitation with nearly identical results. Benzylamine was purchased through Aldrich Chemical Company, Inc. and used without further purification.

7. The product purity was >95% as determined by ${}^1H$ NMR. The checkers analyzed each sample by LC/MS using an ELS detector. For steps 2–6, the mass of the major peak was consistent with the expected mass of the desired product. Product purity as determined by ELS integration of the LC were as follows: product **2**, 89.9%; product **3**, 91.6%; product **4**, 89.8%; product **5**, 90.4%.

8. Acetic anhydride was obtained from Aldrich Chemical Company and used without further purification.

9. Yield of the final product was determined based on the isolated yield of material in the final step from the calculated loading of resin **5** (51.0–75.6%) and for the six-step sequence from the reported manufacturer's loading (29–44%). The acetylated di-β-peptoid **6** exists in solution as a mixture of four conformers, which can be clearly seen by ${}^1H$ NMR at

room temperature in DMSO (Fig. 6.2A). Four signals are seen for the acetyl methyl group (1.96, 2.00, 2.04, and 2.14 ppm) in roughly equal proportions. The remaining signals appear as complicated multiplets. Upon heating to 60°C, the methyl signals broaden and begin to coalesce at 80°C. Reasonably sharp signals were obtained at 125°C (Fig. 6.2B, the temperature limits of our probe) and assured us that we have a single compound rather than an undefined mixture. The high temperature also allowed collection of carbon spectra: ^{13}C NMR (DMSO-d_6, 125°C) δ 20.7, 31.1, 32.5, 42.2. 43.5 (broad s), 50.0 (broad s), 126.7, 128.1, 137.4, 137.7, 169.7, 170.4, 171.9.

10. During our initial test runs on the Argonaut Nautilus 2400 synthesizer, the reaction vessels were not adequately cooled after completion of the Michael addition reaction. Because the reaction vessels are contained in a small cabinet, they did not cool quickly enough to provide a wash cycle at room temperature even though the vessel heater was turned off. As a result, some of the initially formed addition product underwent a retro-Michael addition during the wash cycle to provide the acrylamide **3** and the desired product **9**. Any amine that is released into the wash solution as a result of the retro-Michael addition is washed away, leaving a mixture of products on the resin for eventual cleavage. This had not been observed in reactions carried out in manual reactors, because they were cooled before addition of wash solvents. The Nautilus program was adjusted to allow cooling of the vessels before draining the reaction mixture and carrying out the wash cycle. The vessels were equilibrated to 20°C before the wash cycle to reduce the possibility of a retro-Michael addition and loss of desired product **9**.

11. The excess *n*-dodecylamine from the reaction vessel **k** is not appreciably soluble in the regular wash solvents, particularly at 20°C. Treatment with 10% aqueous acetic acid before the

A- Collected at probe T = 25°C

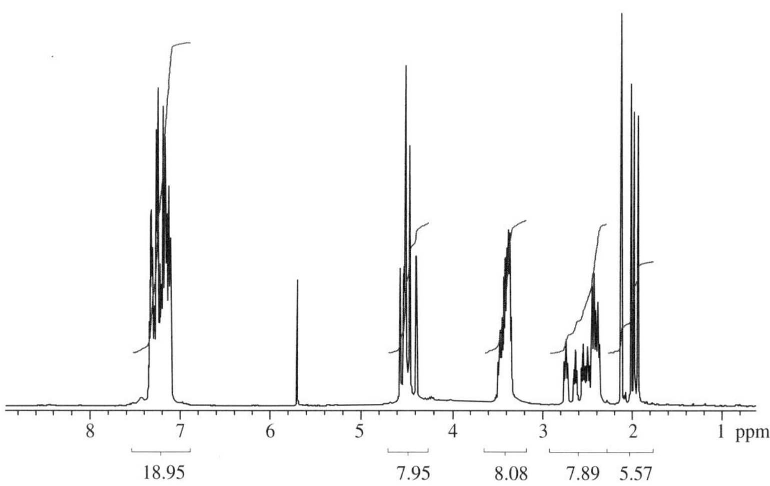

B- Collected at probe T = 110°C

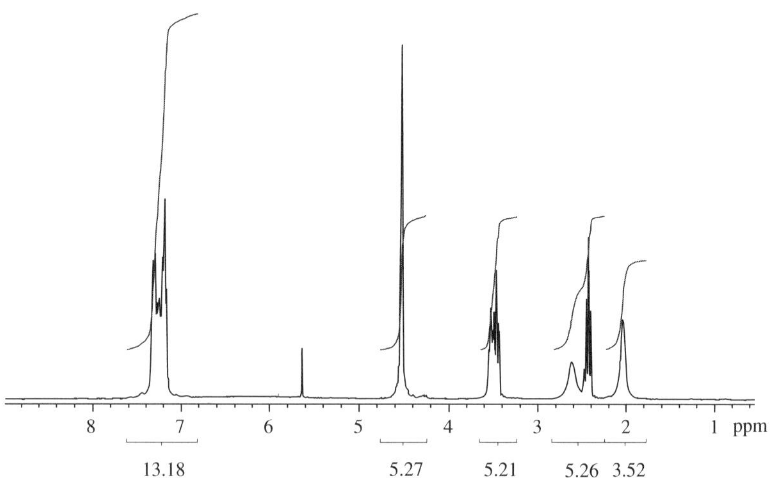

Figure 6.2. ^{1}H NMR spectra of *N*-acetyl-*N*-benzyl-β-alanine-*N*-benzyl-β-alanine **6** in DMSO-d_6 at (**A**) room temperature and (**B**) at 110°C.

wash sequence removed any excess amine without causing premature cleavage of the resin. Even with the acid wash, the checkers observed an impurity in **9k**, presumably dodecylamine, by HPLC using ELS detection, which was not significantly visible by UV at 210 or 254. The ability to remove the *n*-dodecylamine also depends on the efficiency of the wash cycle in the automated synthesis device and may differ in the two instruments employed.

12. Although it is possible to carry out the cleavage of the resins directly on the Nautilus, the delivery of the HMDS standard solution is not sufficiently accurate to allow determination of loadings by ^{1}H NMR. After removal of the vessels from the reaction module, 1.00 mL of the HMDS solution was carefully added by gas tight syringe.

13. The checkers prepared compounds **9a−k** using an ACT 496, in a 4 × 4 10-mL Teflon block with controls occupying the extra four cells. Samples were concentrated to dryness after cleavage, which precluded the use of direct cleavage ^{1}H NMR for determination of purity and yield. Recovered weight and coupled LC/MS analysis was used to determine yield and purity, with results nearly equivalent to the NMR method. Results from this set of parallel reactions were as follows: (compound: weight, % yield, % purity by LC/MS), **9a**: 16 mg, 85%, > 95%; **9b**: 16.5 mg, 81%, > 95%; **9c**: 18 mg, 86%, > 95%; **9d**: 12 mg, 70%, 93%; **9e**: 16 mg, 93%, > 95%; **9f**: 15 mg, 86%, > 95%; **9g**: 14 mg, 80%, 90%; **9h**: 24 mg, 94%, 83%; **9i**: 14 mg, 78%, 90%; **9j**: 16 mg, 73%, 85%; **9k**: 18.3 mg, 73%, 85%; **9l**: no desired product.

14. Chromatographic analysis was obtained using reverse phase HPLC: Zorbax C18, column dimensions, 4.6 mm inner diameter × 10 cm; mobile phase of CH_3CN/H_2O containing 0.1% TFA; gradient profile: 30% CH_3CN/H_2O for 0.5 min, 30% to 100% CH_3CN over 4.5 min, 100% CH_3CN for 2 min.,

total program, 7 min.; flow rate, 2 mL/min.; UV detection at 210 nm.

DISCUSSION

Solid-phase synthesis is the most convenient method for preparation of oligomeric *N*-substituted β-aminopropionic acids or β-peptoids.[1] The acrylate and acrylamide resins are reactive toward a wide variety of primary amines, allowing introduction of a diverse set of substituents.[1,2] Use of primary amines is essential for chain extension by acylation with acryloyl chloride, although secondary amines can be used as a chain-terminating step for the amine end of an oligomer. β-Peptoids can be prepared by standard peptide couplings of *N*-substituted β-amino acids, however this approach requires the preparation of each of the β-amino acids before solid-phase synthesis. The solid-phase approach also eliminates the formation of bis-addition products (addition of two equivalents of acrylate or acrylamide to the amine); a common side product of solution phase synthesis.[3] Standard coupling of Fmoc amino acids is compatible with the solid-phase procedure for preparation of oligomeric β-aminopropionic acids, as previously shown for inclusion of Fmoc β-alanine and nipecotic acid in a trimer series,[1] allowing the formation of "mixed peptide or peptoid" chains. Reaction of acrylate resins that are not TFA cleavable with secondary amines has been investigated as a means of preparing tertiary amines by Michael addition of the amine, alkylation, and Hoffman elimination from the resin.[4] In this case, the acrylate resin is used as a linker, and the three-carbon unit is not incorporated in the final product.

Addition of amines **8a−k** to acrylamide **3** resulted in good yields (71−92%) of di-β-peptoids **9a−k**, with conversions of the starting resin being >95% in all cases except **9i**, **j**, and **l** (Table 6.1). The α-branched amines typically required longer reaction times for completion. Because double treatments were used in these studies, good conversions were seen even for **9e**, **g**, and **i**. However, for cyclopropylamine adduct **9i**, the presence of unreacted acrylamide from resin **3** was detected by both NMR and HPLC as a 6–10% impurity. Anilines such as **8l** were found to be unreactive toward either acrylate **1** or acrylamide resin **3**. Substitution of the resin-bound acrylate or acrylamide double bond with simple alkyl groups led to little or no reaction with amines. Therefore, the preparation of oligomers having substitution along the carbon backbone are not readily available by this route. Oligomers of substituted β-amino acids can be prepared by carbon elongation of α-amino acids, and coupling of the resultant β-amino acids to afford substitution on the carbon backbone.[5]

Because amines **8a–k** can be added to either the acrylate **1** or acrylamide **3**, it is possible to prepare a set of 121 dimers from the set of eleven amines (11 × 11). Alternative capping groups (R_3) can be added to the resin-bound dimers to increase the number of library members. In addition, the carboxylic acid obtained after cleavage can be esterified to provide additional modification of the final components by solution phase chemistry.[1] It is readily apparent that with the alternate introduction of acrylic acid and amines, it is possible to build large libraries of β-peptoids by solid-phase synthesis from readily available starting materials. The conditions employed are compatible with standard Fmoc coupling procedures, allowing the incorporation of an *N*-substituted β-alanine in place of a natural amino acid in solid-phase peptide synthesis. Using the solid-phase approach to include *N*-substituted β-alanines in larger peptides creates truly limitless possibilities for the synthesis of new libraries.

WASTE DISPOSAL INFORMATION

All toxic materials were disposed of in accordance with *Prudent Practices for Disposal of Chemicals from Laboratories*, National Academy Press; Washington, D.C., 1983.

REFERENCES

1. Hamper, B. C.; Kolodziej, S. A.; Scates, A. M. et al. *J. Org. Chem.* **1998**, *63*, 708.

2. Kolodziej, S. A.; Hamper, B. C. *Tetrahedron Lett.* **1996**, *37*, 5277.

3. Zilkha, A.; Rachman, E. S.; Rivlin, J. *J. Org. Chem.* **1961**, *26*, 376 and Stork, G.; McElvain, S. M. *J. Am. Chem. Soc.* **1947**, *69*, 971.

4. Ouyang, X.; Armstrong, R. W.; Murphy, M. M. *J. Org. Chem.* **1998**, *63*, 1027 and Brown, A. R.; Rees, D. C.; Rankovic, Z.; Morphy, J. R. *J. Am. Chem. Soc.* **1997**, *119*, 3288.

5. Appella, D. H.; Christianson, L. A.; Klein, D. A. et al. *Nature* **1997**, *387*, 381 and Seebach, D.; Overhand, M.; Kuhnle, F. N. M. et al. *Helv. Chim. Acta* **1996**, *79*, 913.

CHAPTER SEVEN

SOLID-PHASE SYNTHESIS OF BENZOXAZOLES VIA MITSUNOBU REACTION

Submitted by **FENGJIANG WANG** and **JAMES R. HAUSKE**

Department of Drug Discovery, Sepracor Inc., 111 Locke Drive, Marlborough, MA, USA 01752

Checked by TERRANCE CLAYTON and R. ALAN CHRUSCIEL

Pharmacia & UpJohn, 7223-209-613, 301 Henrietta Street, Kalamazoo, MI, USA 49007-4940

LIBRARY SYNTHESIS ROUTE

BUILDING BLOCKS

Diamines	dicarboxylic anhydrides	2-aminophenols

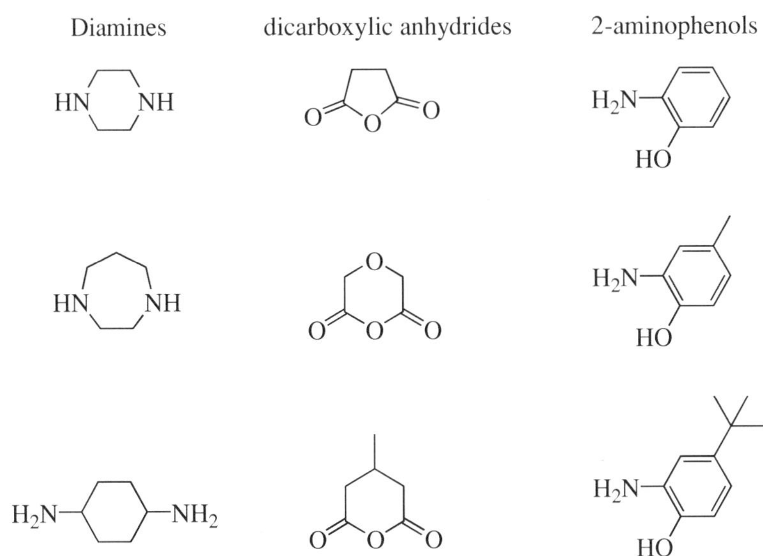

PROCEDURE

A TYPICAL PROCEDURE FOR THE PREPARATION OF INDIVIDUAL BENZOXAZOLE 4

Preparation of Carboxylfunctionalized Resin 1

To the Wang resin (100 mg, 0.070 mmol) in a 3-mL polypropylene filtration tube with polyethylene frit was added 1 mL 0.4 N CDI in anhydrous THF (note 1), capped with a yellow polyethylene cap, and shaken at room temperature for 6 h (note 2). The resin was thoroughly washed with CH_2Cl_2 (3 × 1 mL) and THF (3 × 1 mL) to remove the excess CDI and then treated with 1 mL 0.4 N piperazine in THF at room temperature for 15 h. The resulting resin was washed with DMF (3 × 1 mL), MeOH (4 × 1 mL), and CH_2Cl_2 (4 × 1 mL) and dried *in vacuo*. To the aminofunctionalized resin was added 1 mL 0.4 N succinic anhydride in pyridine/CH_2Cl_2 (v/v = 1:1) and 5 mg DMAP, and the resulting slurry was shaken at room temperature for 4 h. The resulting carboxylfunctionalized **resin 1** was washed with DMF (3 × 1 mL), MeOH (4 × 1 mL), and CH_2Cl_2 (4 × 1 mL) and dried *in vacuo*.

Preparation of Benzoxazole 4

To **resin 1** (0.070 mmol) was added PyBOP (182 mg, 0.35 mmol) and 2-aminophenol (38 mg, 0.35 mmol) in 1 mL DMF, followed by *N*-methylmorpholine (NMM) (38 µL, 0.35 mmol). The mixture was shaken at room temperature for 3 h. The resulting **resin 2** was washed extensively with DMF (3 × 1 mL), MeOH (4 × 1 mL), and CH_2Cl_2 (4 × 1 mL) and dried *in vacuo*. To the mixture of **resin 2** and Ph_3P (92 mg, 0.35 mmol) in 1 mL anhydrous THF was added diethyl azodicarboxylate (DEAD) (55 µL, 0.35 mmol). The reaction mixture was shaken at room temperature for 17 h, followed by washing with DMF (3 × 1 mL), MeOH (4 × 1 mL), and CH_2Cl_2 (4 × 1 mL). The resulting **resin 3** was dried *in vacuo*, treated with a solution of 50% TFA in CH_2Cl_2 (1.5 mL) at room temperature (note 3) for 30 min to release the polymer-bound benzoxazole and washed with CH_2Cl_2 (2 × 1 mL). Removal of the volatiles under a stream of nitrogen followed by drying under high vacuum overnight afforded the crude compound **4**, which was submitted to HPLC, mass spectrum, and NMR analyses (notes 4 and 5).

A DIRECTED LIBRARY SYNTHESIS OF BENZOXAZOLES

As described above, a small library containing 27 benzoxazoles was synthesized by using three diamines, three dicarboxylic anhydrides, and three 2-aminophenols (Table 7.1). Wang resin

TABLE 7.1. Synthesis of a Small Benzoxazoles Library, the Yields and Mass Spectra (M+1)$^+$ (note 6)

	1	2	3
A	92% (260)	90% (276)	89% (288)
B	93% (274)	84% (290)	95% (302)
C	88% (316)	87% (332)	89% (344)
D	88% (274)	83% (290)	86% (302)
E	93% (288)	84% (304)	88% (316)

TABLE 7.1. (*Continued*)

	1	2	3
F	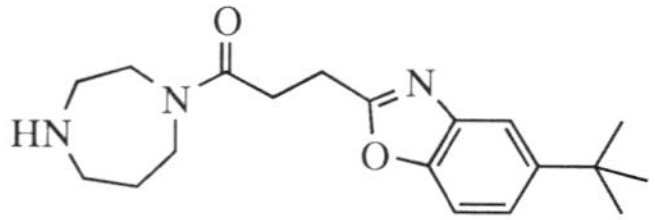 88% (330)	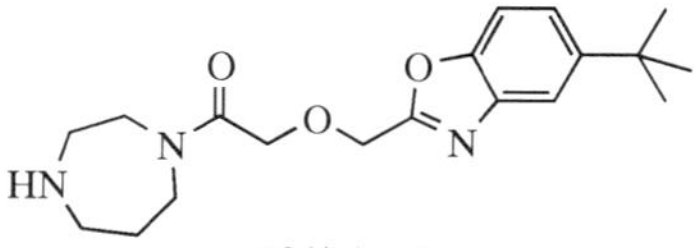 82% (346)	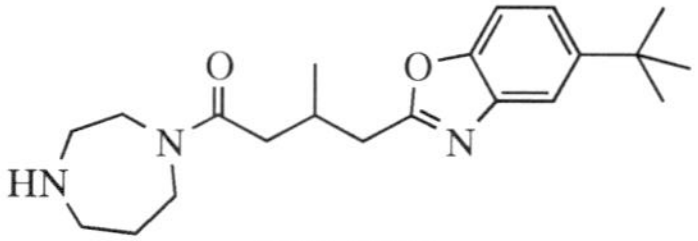 80% (358)
G	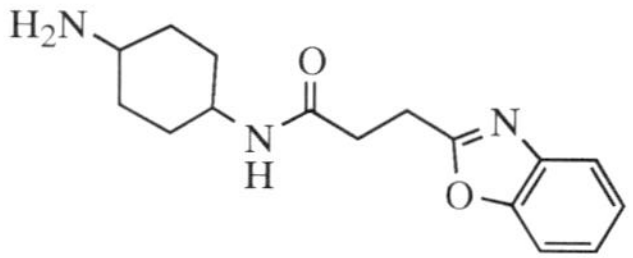 76% (288)	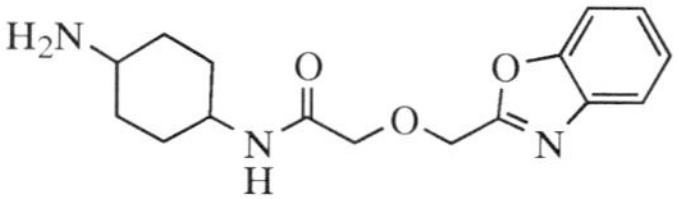 80% (304)	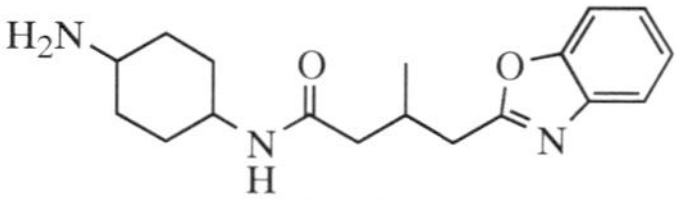 82% (316)
H	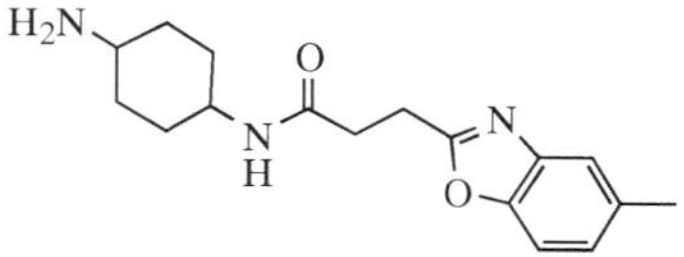 84% (302)	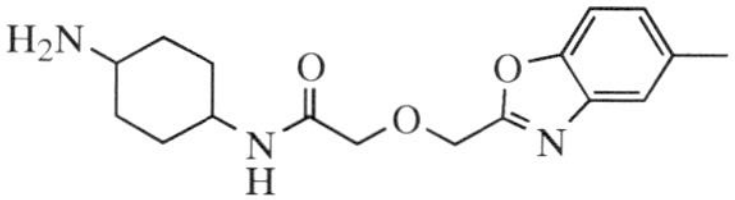 79% (318)	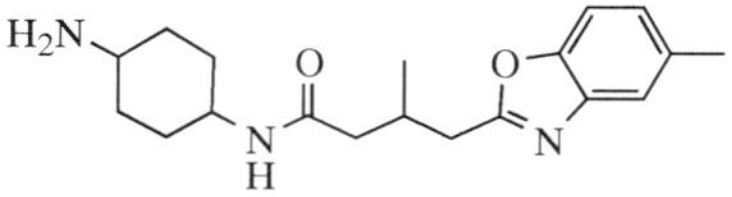 81% (330)
I	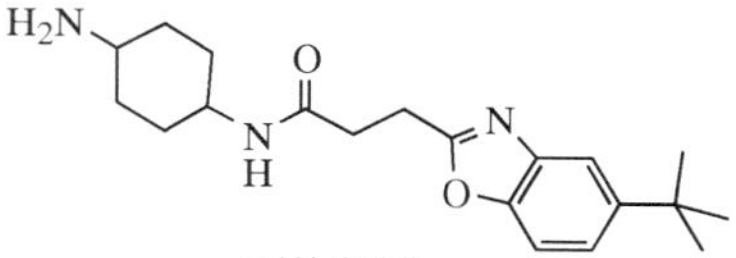 76% (344)	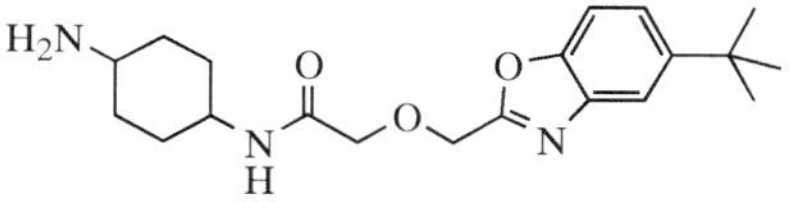 72% (360)	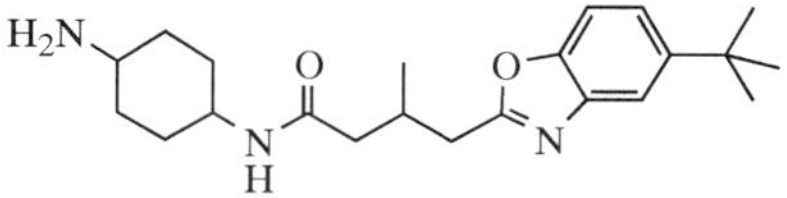 79% (372)

was distributed into twenty-seven 3-mL filtration tubes (100 mg/ tube, 0.070 mmol) followed by adding 1 mL 0.4 N CDI in THF. After shaking at room temperature for 6 h, the resins were washed with CH_2Cl_2 (3 × 1 mL/tube) and THF (3 × 1 mL/tube) to remove the excess CDI. A solution of piperazine in THF (310 mg in 9 mL THF) was dispensed into 9 tubes of row A, row B, and row C at 1 mL/tube; a solution of homopiperazine in THF (361 mg in 9 mL THF) was dispensed into 9 tubes of row D, row E, and row F at 1 mL/tube; and finally, a solution of *trans*-1,4-diaminocyclohexane in THF (411 mg in 9 mL THF) was dispensed into 9 tubes of row G, row H, and row I at 1 mL/ tube. The resulting mixtures were shaken at room temperature for 15 h, and the resins were washed with DMF (3 × 1 mL/tube), MeOH (4 × 1 mL/tube), and CH_2Cl_2 (4 × 1 mL/tube) and dried *in vacuo*. Succinic anhydride solution (360 mg succinic anhydride, 45 mg DMAP, 4.5 mL pyridine, and 4.5 mL CH_2Cl_2) was dispensed into 9 tubes of column 1 at 1 mL/tube; diglycolic anhydride solution (418 mg diglycolic anhydride, 45 mg DMAP, 4.5 mL pyridine, and 4.5 mL CH_2Cl_2) was dispensed into 9 tubes of column 2 at 1 mL/tube; and finally, 3-methylglutaric anhydride solution (461 mg 3-methylglutaric anhydride, 45 mg DMAP, 4.5 mL pyridine, 4.5 mL CH_2Cl_2) was dispensed into 9 tubes of column 3 at 1 mL/tube. The reaction mixtures were agitated at room temperature for 4 h. The resulting carboxylfunctionalized resins (**1**) were then washed with DMF (3 × 1 mL/ tube), MeOH (4 × 1 mL/tube), and CH_2Cl_2 (4 × 1 mL/tube) and dried *in vacuo*.

Next, PyBOP (4.914 g, 9.45 mmol) in 13.5 mL DMF was dispensed into all the reaction tubes at 0.5 mL/tube and 2-aminophenol (344 mg, 3.15 mmol) in 4.5 mL DMF was dispensed into 9 tubes of row A, row D, and row G at 0.5 mL/tube; 2-amino-*p*-cresol (388 mg, 3.15 mmol) in 4.5 mL DMF was dispensed into 9 tubes of row B, row E, and row H at 0.5 mL/tube; and finally, 2-amino-4-*tert*-butylphenol (521 mg, 3.15 mmol) in 4.5 mL DMF was dispensed into 9 tubes of row C, row F, and row I at 0.5 mL/

tube. NMM (38 μL, 0.35 mmol) was then added into each one of the reaction tubes. After agitating at room temperature for 3 h, the resulting 2-amidophenol resins (**2**) were washed with DMF (3 × 1 mL / tube), MeOH (4 × 1 mL / tube), CH_2Cl_2 (4 × 1 mL / tube) and dried *in vacuo*. Triphenylphosphine (2.48 g, 9.45 mmol) in 27 mL anhydrous THF was dispensed into all the reaction tubes at 1 mL / tube followed by addition of DEAD (55 μL / tube, 0.35 mmol). After shaking at room temperature for 17 h, the resulting resins (**3**) were washed with DMF (3 × 1 mL / tube), MeOH (4 × 1 mL / tube), CH_2Cl_2 (4 × 1 mL / tube) and dried *in vacuo*. The resulting resins (**3**) were treated with a solution of 50% TFA in CH_2Cl_2 (1.5 mL / tube) at room temperature for 30 min to release the polymer-bond benzoxazoles (**4**). After washing the resins with CH_2Cl_2 (2 × 1 mL / tube), the volatiles were removed under a stream of nitrogen followed by drying under high vacuum overnight to afford the crude compounds. These compounds were submitted to HPLC, mass spectra, and NMR analyses.

NOTES

1. Wang resin was purchased from Advanced ChemTech (1% DVB, 0.70 mmol / g substitution, 100–200 mash, Cat. # SA5009). Anhydrous tetrahydrofuran (THF), *N,N*-dimethylformamide (DMF), methanol, dichloromethane, pyridine, 1,1′-carbonyldiimidazole (CDI), piperazine, homopiperazine, *trans*-1,4-diaminocyclohexane, 4-(dimethylamino)pyridine (DMAP), succinic anhydride, diglycolic anhydride, 3-methylglutaric anhydride, 2-aminophenol, 2-amino-*p*-cresol, 2-amino-4-*tert*-butylphenol, *N*-methylmorpholine (NMM), triphenylphosphine, diethyl azodicarboxylate (DEAD), and trifluoroacetic acid (TFA) were purchased from Aldrich Chemical Company, Inc. and used without further purification. PyBOP was purchased from Novabiochem.

2. Polypropylene filtration tubes (3 mL) with polyethylene frits were purchased from Supelco (Cat. # 5-7024). The filtration tubes were capped by using a yellow polyethylene cap (custom order from Supelco) for 3-mL filtration tube. The bottom of the tubes was sealed by inserting a female luer plug (Supelco Cat. # 5-7098) into the bottom of the tube. Tubes were horizontally placed on an IKA orbital shaker (model KS250) and shaken at 200 rpm. All reactions were conducted without precaution to exclude atmospheric oxygen or moisture. The checkers capped the filtration tubes using polyethylene caps from Baxter (Cat. # T-1226-32). The bottom of the tubes were sealed by inserting them into inverted septa of appropriate diameter (Aldrich Cat. # Z16,725-8). Shaking was effected using a LabLine orbital shaker (model 4626). Tubes were placed in a horizontal position and shaken at 110 rpm. The checkers observed that yields of the key Mitsunobu reaction were improved when the reagents were added under a nitrogen atmosphere within a glove bag (Aldrich Cat. # Z11,835-4).

3. The bottom of the filtration tube was equipped with a one-way stopcock (Alltech Cat. # 213112), which was closed to prevent drainage. After 30 min, the stopcock opened, the cleavage solution drained into a test tube, and the resin was washed with CH_2Cl_2. The checkers cleaved the samples from resin by adding a solution of 50% TFA in CH_2Cl_2 (1.5 mL) at room temperature to the filtration tube equipped (on the bottom) with a disposable flow control valve line (Supelco Cat. # 5-7059), which was further clamped to prevent drainage. After 30 min, the clamp was removed, the cleavage solution drained, and the resin washed with CH_2Cl_2 (2 × 1 mL).

4. ^{1}H NMR spectra were recorded on a Varian Inova NMR 300 spectrometer operating at 300 MHz. ESI Mass spectra were obtained on a Micromass Platform LC Mass Spectrometer. The HPLC analyses were performed on a Hewlett Packard

1100 system equipped with a ZORBAX Rx-C18, 4.6-mm inner diameter × 25 cm (5 μM) column monitoring at both 214 nM and 254 nM. Elution was performed at a flow rate of 1.0 mL / min with 0.05% aqueous TFA and a linear gradient of 5–100% acetonitrile containing 0.05% TFA over 10 min. The checkers recorded ^{1}H NMR spectra on a Bruker Avance DPX 300 spectrometer operating at 300 MHz. Electrospray mass spectra were obtained using a Micromass Platform II spectrometer. HPLC chromatograms were performed on a Gilson 712 instrument equipped with a 4.6 × 250 mm, 10 μM, C-18 Vydac 218tp54 column monitoring at 210 nM. Elution was performed at a flow rate of 1.5 mL / min with 0.1% aqueous TFA and a linear gradient of 10–90% acetonitrile containing 0.07% TFA over 18 min.

5. The structure of this individual compound **4** is as same as the structure of **A-1** in Table 7.1. ^{1}H NMR (DMSO-d_6) 2.99 (t, J = 7.5 Hz, 2H), 3.07 (bs, 2H), 3.16 (t, J = 6.3 Hz, 4H), 3.63–3.66 (m, 2H), 3.72–3.75 (m, 2H), 7.27–7.44 (m, 2H), 7.62–7.65 (m, 2H), 9.06 (bs, 2H). MS (EI) m/z 260 (MH)$^+$.

6. Yields of the products were determined by using the NMR integration of a sample containing 2-methylbenzoxazole (8.3 μL, 0.07 mmol) as an internal standard in DMSO-d_6, in which the peak of the methyl protons at 2.60 ppm was the standard peak for the comparison with the 2-methylene protons of the crude benzoxazoles. The yields observed by the checkers from the preparation of the directed library are **A-1** (95%), **A-2** (81%), **A-3** (95%)*, **B-1** (94%), **B-2** (85%), **B-3** (95%)*, **C-1** (84%), **C-2** (73%), **C-3** (73%), **D-1** (61%), **D-2** (74%), **D-3** (77%), **E-1** (84%), **E-2** (80%), **E-3** (84%)*, **F-1** (89%), **F-2** (76%), **F-3** (81%)*, **G-1** (63%)*, **G-2** (70%), **G-3** (76%), **H-1** (74%)*, **H-2** (65%), **H-3** (79%), **I-1** (54%)*, **I-2** (64%), **I-3** (69%). Owing to overlapping chemical shifts (integrals) with the standard, those yields with * are approximate. In all cases, the parent ions of the target

compounds were observed by ESI MS. Qualitative analyses of the HPLC chromatograms were consistent with the NMR results.

DISCUSSION

Thermal cyclization with acid catalysts are commonly employed to synthesize benzoxazoles.[1] For example, 2-amidophenols have been treated with PPA or PPE,[2,3] propionic acid,[4] $POCl_3$,[5] and $SOCl_2$[6] at high temperature to give benzoxazoles. It was noted that those conditions were not suitable for solid-phase synthesis, because the polymer support and the linker normally do not survive under such harsh reaction conditions. When we exposed solid-phase linked 2-amidophenols to either $POCl_3$ or $SOCl_2$ with 1 Eq. pyridine in toluene at 80°C, > 50% of the 2-amidophenol was cleaved from solid support in 30 min. The intramolecular dehydrative cyclization of the 2-amidophenol attached to a solid support employing excess of Ph_3P and DEAD in THF proceeded smoothly at room temperature to provide resin-bond benzoxazole. In general, the reaction of **resins 2** under Mitsunobu conditions[7] gave benzoxazoles in high yield and in high purity. With an electron-withdrawing group on the aromatic ring, for example, 4-chloro-2-amidophenol, the yield and the purity of the resulting benzoxazole was adversely effected.[8]

REFERENCES

1. Boyd, G. V. In Katritzky, A. R.; Rees, C. W., eds., *Comprehensive Heterocyclic Chemistry*, vol. 6, part 4B, Pergammon: Oxford, UK 1984, p. 178.

2. Suto, M. J.; Turner, W. R. *Tetrahedron Lett.* **1995**, *36*, 7213.

3. Haugwitz, R. D.; Angel, R. G.; Jacobs, G. A. et al. *J. Med. Chem.* **1982**, *25*, 969.

4. Nestor, J. J.; Norner, B. L.; Ho, T. L. et al. *J. Med. Chem.* **1984**, *27*, 320.

5. Orjales, A.; Bordell, M.; Rubio, V. *J. Heterocyclic Chem.* **1995**, *32*, 707.

6. Stack, J. G.; Curran, D. P.; Geib, S. V. et al. *J. Am. Chem. Soc.* **1992**, *114*, 7007.

7. Mitsunobu, O. *Synthesis* **1981**, 1. and Hughes, D. L. *Org. React.* **1992**, *42*, 335.

8. Wang, F.; Hauske, J. R. *Tetrahedron Lett.* **1997**, *38*, 6529.

CHAPTER EIGHT

N-FMOC-AMINOOXY-2-CHLOROTRITYL POLYSTYRENE RESIN FOR HIGH THROUGHPUT SYNTHESIS OF HYDROXAMIC ACIDS

Submitted by WENG C. CHAN*, SARAH L. MELLOR, and
GAIL E. ATKINSON

School of Pharmaceutical Sciences, University of Nottingham,
University Park, Nottingham, England, NG7 2RD

Checked by EDWARD L. FRITZEN* and
DOUGLAS J. STAPLES[†]

*Combinatorial and Medicinal Chemistry and [†]Research Operations;
Pharmacia Corp., 7000 Portage Road,
Kalamazoo MI, USA 49001

REACTION SCHEME

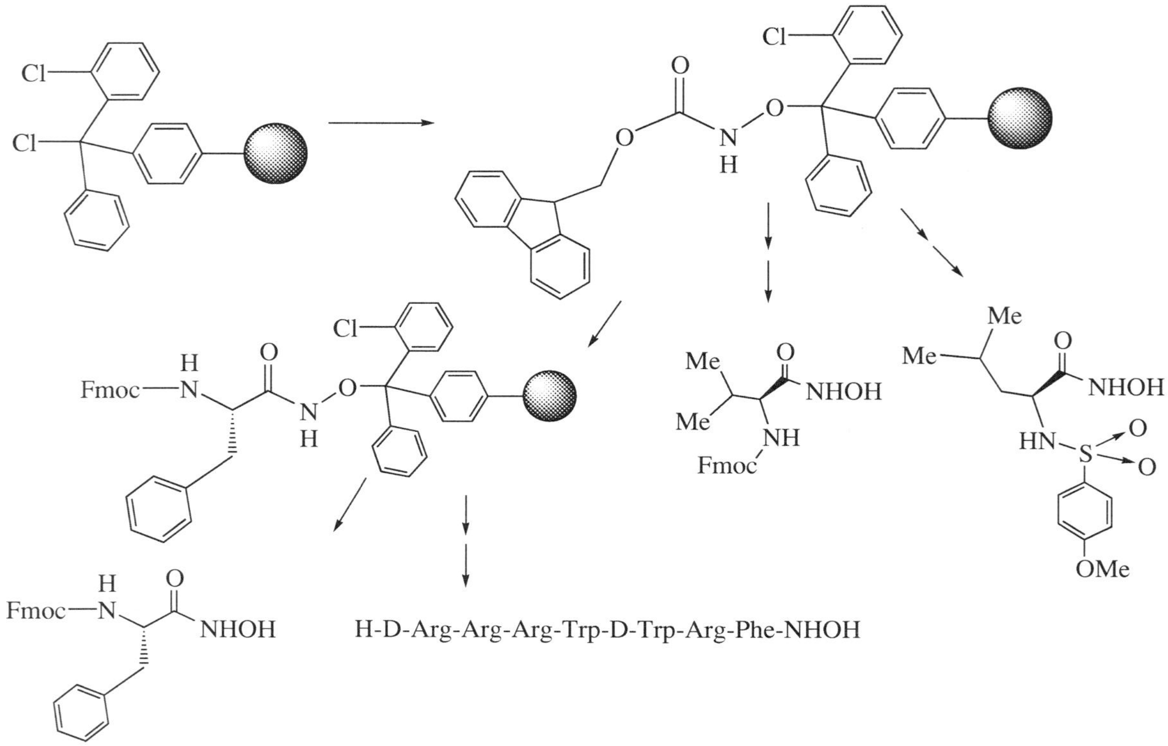

PROCEDURES

Abbreviations

DCM: dichloromethane.

DIEA: *N,N*-diisopropylethylamine.

DMF: *N,N*-dimethylformamide.

HATU: *N*-[(dimethylamino)-1*H*-1,2,3-triazolo[4,5-*b*]pyridin-1-ylmethylene]-*N*-methylmethanaminium hexafluorophosphate *N*-oxide (also known as *O*-(7-azabenzotriazol-1-yl)-1,1,3,3-tetramethyluronium hexafluorophosphate).

HOAt: 1-hydroxy-7-azabenzotriazole.

HOBt: 1-hydroxybenzotriazole.

TBTU: *N*-[(1*H*-benzotriazol-1-yl)(dimethylamino)methylene]-*N*-methylmethanaminium tetrafluoroborate (*O*-(benzotriazol-1-yl)-1,1,3,3-tetramethyluronium tetrafluoroborate).

TFA: trifluoroacetic acid

RP-HPLC methods

Column: Hypersil Pep5-C18 (4.6 × 150 mm); solvent A: 0.06% aqueous TFA; solvent B: 0.06% TFA in 90% aqueous acetonitrile; flow rate: 1.20 mL min^{-1}; effluent monitored at 220 nm.

*Linear elution gradient **G1:*** 50–100% B in 20 min.

*Linear elution gradient **G2:*** 20–60% B in 25 min.

N-(9-Fluorenylmethoxycarbonyl)hydroxylamine[1]

An aqueous solution of sodium hydrogen carbonate (1.85 g, 22.0 mmol, 20 mL) followed by ethyl acetate (ca. 40 mL) was added to hydroxylamine·hydrochloride (695 mg, 10.0 mmol) in a

100-mL round-bottom flask. The resultant biphasic mixture was stirred and cooled to 5°C. Fmoc-Cl (2.59 g, 10.0 mmol), dissolved in ethyl acetate (10 mL), was then added dropwise to the rapidly stirred biphasic hydroxylamine solution over a period of 30 min (note 1). After the addition, the mixture was allowed to reach ambient temperature and vigorously stirred for a further 3–4 h. The progress of the reaction was monitored by silica-TLC (ethyl acetate: hexane (1:1), Fmoc-NHOH $R_f = 0.14$). The reaction mixture was then separated and the organic phase washed with saturated aqueous potassium hydrogen sulfate (3 × 40 mL) and saturated aqueous sodium chloride (2 × 40 mL). The organic extract was dried over anhydrous magnesium sulfate, filtered, and evaporated to dryness in vacuo to afford, after trituration with hexane, Fmoc-NHOH (2.295 g, 90%) as a white crystalline solid. The product obtained is of high purity, but may be further purified by careful recrystallization from ethyl acetate:hexane.

M.p. 164.5-167.5°C. Electrospray (ES)-MS, m/z 278.3 (M + Na$^+$; calculated, 278.08).

δ_H (250 MHz, CDCl$_3$) 4.21 (1H, t, J 6.9 Hz, Fmoc CH), 4.32 (2H, d, J 6.7 Hz, Fmoc CH$_2$), 7.28–7.43, 7.68, 7.86 (8H, m, Fmoc Ar. CHs), 8.77 (1H, s, NH), 9.75 (1H, br s, OH).

δ_C (62.90 MHz, CDCl$_3$) 47.49 (Fmoc CH), 66.44 (Fmoc CH$_2$), 120.86, 126.00, 127.85, 128.53 (Fmoc Ar. CH), 141.57, 144.52 (Fmoc Ar. C), 158.46 (C=O).

N-(9-Fluorenylmethoxycarbonyl)aminooxy-2-chlorotrityl Polystyrene Resin[1]

2-Chlorotrityl chloride polystyrene[2] (84 mg, 0.1 mmol, 1.2 mmol g^{-1}; 1% DVB, 100–200 or 200–400 mesh; CN Biosciences UK Ltd.) was pre-swollen in dry DCM (3 mL; note 2) for 10 min. *N*-Fmoc hydroxylamine (51 mg, 0.2 mmol) followed by DIEA (35 µL, 0.2 mmol) was added, and the reaction mixture (note 3) was stirred at room temperature for 48 h under nitrogen atmosphere. Methanol (0.1 mL) was then added and the mixture stirred for a further 30 min. The resin was then collected using a Buchner funnel, and successively washed with DMF (30 mL), dichloromethane (25 mL), and hexane (5 mL) (note 4), and dried in vacuo over potassium hydroxide pellets for 24 h.

Amount of resin product recovered 87 mg.

Fmoc-substitution (note 5) 0.94 mmol g^{-1}, 92% efficiency (typically 0.8–0.9 mmol g^{-1}; note 6); RP-HPLC analysis (**G1**) of product obtained following acidolytic treatment (5% TFA in CH_2Cl_2, 5 min) showed the exclusive presence of Fmoc-NHOH.

v_{max} (KBr) 1701 (s, C = O), 1445, 1530 and 1554 (m, polystyrene) cm^{-1}.

N-(9-Fluorenylmethoxycarbonyl)phenylalanyl Hydroxamic Acid

N-Fmoc-aminooxy-2-chlorotrityl polystyrene (212 mg, 0.95 mmol g^{-1}, 0.2 mmol) was placed in a reaction column (1.0 cm diameter; alternatively, an appropriate reaction vessel can be used, e.g., Quest 210 synthesizer 5-mL reaction vessel) and preswollen in DCM:DMF (1:1, 3 mL) for 24 h (note 4). The resin was then washed with DMF (10 min, 2.5 mL min^{-1}) and Fmoc-deprotected by treatment with 20% v/v piperidine in DMF (10 min, 2.5 mL min^{-1}). The resin was then washed with DMF (10 min, 2.5 mL min^{-1}), after which excess DMF was removed.

Fmoc-Phe-OH (310 mg, 0.8 mmol), HOAt (108 mg, 0.8 mmol) and HATU[3] (310 mg, 0.8 mmol) were dissolved in DMF (2.0 mL), and DIEA (280 µL, 1.6 mmol) was then added. After ca. 1 min, the mixture was added to the resin and the reaction suspension gently agitated at room temperature for 24 h (note 7). The resin was then washed with DMF (10 min, 2.5 mL min^{-1}); collected in a Buchner funnel; and successively washed with DMF (10 mL), DCM (20 mL), and hexane (5 mL), and dried in vacuo overnight.

Amount of resin product recovered 232 mg.

Fmoc-substitution (note 5) 0.81 mmol g^{-1}, 97% acylation efficiency.

The derivatized resin product (100 mg, 0.08 mmol) was suspended in DCM (6 mL) for 30 min, after which 0.06 mL TFA was added and the resultant suspension was gently stirred for 15 min at ambient temperature. The suspension was filtered, the spent resin was washed with DCM (5 mL) and DCM:MeOH (1:1, 5 mL), and the filtrate was evaporated to dryness in vacuo to give the title compound (29 mg, 90%) as white crystalline solid. RP-HPLC analysis (**G1**) showed the exclusive presence ($> 98\%$) of Fmoc-Phe-NHOH ($R_t = 6.8$ min).

m/z (ES($+$)) calculated, 403.17 (MH$^+$), observed, 403.4 (MH$^+$).

δ_H (250 MHz, [^{2}H]$_6$-DMSO) 2.88 (1H, m, Phe C$^\beta$H), 3.98–4.21 (4H, m, Phe C$^\alpha$H & Fmoc CHCH$_2$), 7.13–7.43, 7.66, 7.87 (13H, m, Phe Ar. CHs & Fmoc Ar. CHs), 7.76 (1H, d, J 8.7 Hz, Phe N$^\alpha$H), 8.92 (1H, br s, NH), 10.75 (1H, br s, OH).

N-(9-Fluorenylmethoxycarbonyl)valinyl Hydroxamic Acid

N-Fmoc-aminooxy-2-chlorotrityl polystyrene (115 mg, 1.00 mmol g^{-1}, 0.115 mmol) was treated as outlined above and Fmoc-deprotected using 20% v/v piperidine in DMF (10 min, 2.5 mL min^{-1}). The resin was then washed with DMF (10 min, 2.5 mL min^{-1}) after which excess DMF was removed. Fmoc-Val-OH (204 mg, 0.6 mmol), HOAt (81 mg, 0.6 mmol) and HATU[3] (232 mg, 0.6 mmol) were dissolved in DMF (1.2 mL) and DIEA

(209 μl, 1.2 mmol) then added. After ca. 1 min, the mixture was added to the resin and the reaction suspension gently agitated at room temperature for 24 h (note 7). The resin was then washed with DMF (10 min, 2.5 mL min^{-1}), collected in a Buchner funnel, and successively washed with DMF (10 mL), DCM (20 mL) and hexane (5 mL), and dried in vacuo overnight.

Amount of resin product recovered was 124 mg.

Fmoc-substitution (note 5) 0.78 mmol g^{-1}, 87% acylation efficiency.

The derivatised resin product was suspended in DCM (6 mL) for 30 min, after which TFA (0.06 mL) was added and the resultant suspension was gently stirred for 15 min at ambient temperature. The suspension was filtered, the spent resin washed with DCM (5 mL) and DCM:MeOH (1:1, 5 mL), and the combined filtrate was evaporated to dryness in vacuo to afford the title compound (25 mg, 73%) as a white crystalline solid. RP-HPLC analysis (**G1**) showed the exclusive presence (>98%) of Fmoc-Val-NHOH (R_t = 5.1 min).

m/z (ES(+)) calculated 355.17 (MH$^+$), observed, 355.0 (MH$^+$), 377.2 (M+Na$^+$).

δ_H (250 MHz, [^{2}H]$_6$-DMSO) 0.87 (3H, d, J 6.97 Hz, Val C$^\gamma$H$_3$), 0.91 (3H, d, J 6.87 Hz, Val C$^\gamma$H$_3$), 1.94 (1H, m, Val C$^\beta$H), 3.66 (1H, t, J 8.78 Hz, Fmoc CH), 4.17–4.33 (3H, m, Val C$^\alpha$H & Fmoc CH$_2$), 7.29–7.45, 7.76, 7.87 (9H, m, NH & Fmoc Ar. CHs), 7.53 (1H, d, J 9.0 Hz, Val N$^\alpha$H), 10.68 (1H, br s, OH).

N-(4-Methoxybenzenesulphonyl)leucyl Hydroxamic Acid

N-Fmoc-aminooxy-2-chlorotrityl polystyrene (100 mg, 1.00 mmol g^{-1}, 0.1 mmol) was treated as outlined above and Fmoc-deprotected using 20% v/v piperidine in DMF (10 min, 2.5 mL min^{-1}). The resin was then washed with DMF (10 min, 2.5 mL min^{-1}) after which excess DMF was removed. Fmoc-Leu-OH (212 mg, 0.6 mmol), HOAt (81 mg, 0.6 mmol), and HATU^3 (232 mg, 0.6 mmol) were dissolved in DMF (1.2 mL); and DIEA was (209 μL, 1.2 mmol) then added. After ca. 1 min, the mixture was added to the resin and the reaction suspension gently agitated at room temperature for 24 h (note 7). The resin was then washed with DMF (10 min, 2.5 mL min^{-1}) and Fmoc deprotected using 20% v/v piperidine in DMF (7 min, 2.5 mL min^{-1}). The resin was then washed with DMF (10 min, 2.5 mL min^{-1}), after which the excess DMF was removed. A solution of 4-methoxysulphonyl chloride (83 mg, 0.4 mmol) in DMF (1 mL) was added to the resin, followed by DIEA (26 μL, 0.15 mmol). The resultant suspension was gently agitated at room temperature for 24 h. The resin was then washed with DMF (10 min, 2.5 mL min^{-1}); collected in a Buchner funnel; and successively washed with DMF (10 mL), DCM (20 mL), and hexane (5 mL); and dried in vacuo overnight. The amount of resin product recovered was 105 mg.

The derivatized resin product was suspended in DCM (6 mL) for 30 min, after which TFA (0.06 mL) was added; the resultant suspension was stirred for 10–15 min at ambient temperature (note 8). The suspension was filtered, the spent resin was washed with DCM (5 mL) and DCM:MeOH (1:1, 5 mL), and the combined filtrate was evaporated to dryness in vacuo to afford the title compound (28 mg, 90%). RP-HPLC analysis (**G2**) showed predominantly (>90%) *N*-(4-methoxy-benzenesulpho-nyl)-leucyl hydroxamic acid (R_t = 10.4 min).

m/z (ES(+)) calculated, 317.12 (MH$^+$); observed, 317.3 (MH$^+$).

δ_H (250 MHz, CDCl$_3$:[^{2}H]$_6$-DMSO) 0.70 (3H, d, J 6.3 Hz, Leu CH$_3$), 0.83 (3H, d, J 6.4 Hz, Leu CH$_3$), 1.37–1.60 (3H, m, Leu C$^\beta$H$_2$ and C$^\gamma$H), 3.74 (1H, m, Leu C$^\alpha$H), 3.87 (3H, s,

OCH$_3$), 6.83 (1H, d, J 8.3 Hz, Leu NH), 6.96 (2H, d, J 8.8 Hz, Ar Hs), 7.46 (1H, s, NH), 7.79 (2H, d, J 8.8 Hz, Ar Hs).

H-D-Arg-Arg-Arg-Trp-D-Trp-Arg-Phe-NHOH

N-(Fmoc-Phe)-aminooxy-2-chlorotrityl polystyrene (88 mg, 0.60 mmol g^{-1}, 0.0528 mmol), placed in a reaction column (note 9) was left in DMF (1 mL) for 18 h and then Fmoc-deprotected using 20% v/v piperidine in DMF (10 min, 2.5 mL min^{-1}). The resin was then washed with DMF (10 min, 2.5 mL min^{-1}), and the peptide sequence H-D-Arg(Pmc)-Arg(Pmc)-Arg(Pmc)-Trp(Boc)-D-Trp(Boc)-Arg(Pmc)- was assembled using the automated Milli-Gen PepSynthesizer 9050 (note 9).

Sequential acylation reactions were carried out at ambient temperature for 1.5 h using a DMF solution (1.3 mL) of the appropriate *N*-Fmoc–protected amino acids [Fmoc-Arg/D-Arg(Pmc)-OH, 265 mg; Fmoc-Trp/D-Trp(Boc)-OH, 211 mg; 0.4 mmol) and then carboxyl activated using TBTU (154 mg, 0.4 mmol), HOBt (54 mg, 0.4 mmol), and DIEA (140 μL, 0.8 mmol). Repetitive N^α-Fmoc deprotection was achieved using 20% v/v piperidine in DMF (6 min, 2.5 mL min^{-1}).

The assembled N^α-Fmoc-deprotected peptidyl resin was collected in a Buchner funnel; washed with DMF (10 mL), DCM (20 mL), and MeOH (5 mL); and dried in vacuo overnight. The amount of resin product recovered 162 mg (0.0433 mmol).

The resin product was suspended in TFA (9 mL), into which was immediately added water (0.45 mL), 1,2-ethanedithiol (0.45 mL), and triisopropylsilane (0.1 mL). The mixture was left, with occasional agitation, at 30°C for 4 h. The suspension was then filtered, the spent resin washed with TFA (3 × 1 mL) and the combined filtrate was evaporated to dryness in vacuo. The residual material was then triturated with diethyl ether (10 mL) to give a white solid, which was filtered, washed with diethyl ether (3 × 10 mL), and dried in vacuo to afford the title compound

(47 mg, 97%) as a white solid. Based upon RP-HPLC analysis (**G2**), the purity (note 10) is estimated to be 90%.

$R_t = 10.0$ min; m/z (ES(+)) calculated, 1177.65 (MH$^+$); observed, 1177.9 (MH$^+$).

NOTES

1. The use of an excess of Fmoc-Cl (1.5 Eq.) and/or stronger basic conditions typically promote significant formation of the undesired *bis*-protected compound, *N,O-bis*-Fmoc-hydroxylamine (m.p. 159.5–161°C; ES-MS, m/z 478.4 (MH$^+$; calculated, 478.17); silica-TLC (ethyl acetate:hexane, 1:1) $R_f = 0.64$.

2. DCM is redistilled from calcium hydride and stored over molecular sieve. This reaction can be carried out in an oven-dried round-bottomed flask (10 mL) or using the Quest 210 semiautomated synthesizer 5-mL reaction vessels.

3. Fmoc-NHOH is generally not very soluble in DCM; freshly redistilled THF (ca. 1 mL) may be added to aid dissolution.

4. This causes the resin to shrink and aids in the handling of resin material. As a result, the dried resin product must be preswollen in DCM:DMF (1:1), DCM:THF (1:1), or DCM for 24 h before use for solid-phase chemistry.

5. The resin substitution level is based on spectrophotometric determination of the Fmoc-derived chromophore liberated upon treatment with 20% piperidine/DMF using $\epsilon_{290\ nm} = 5253$ M^{-1} cm^{-1}, which was used to calculate the percent efficiency.

6. The Checkers found that the condensation reaction was variable and could range from 36 to 54% Fmoc-substitution

levels. More consistent results (57–78%) were obtained when the reaction was carried out using 5 Eq. Fmoc-NHOH in the presence of 5 Eq. DIEA. Moreover, this alternative approach was found to be reliable when the reaction was performed on a larger scale (1.25 mmol); the resin product gave a Fmoc substitution of 0.70 mmol g^{-1}.

7. Owing to steric hindrance, the acylation reaction must be carried out using a large excess (4–10 Eq.) of the activated acid and for an extended period. In some cases, repeat acylation is recommended. Acylation has also been successfully carried out using Fmoc–amino acid fluorides (e.g., Fmoc-Phe-F[4], 4 Eq. in the presence of DIEA, 1.1 Eq.; 18 h; >98% acylation efficiency). While acylation with unhindered activated carboxylic acids are achieved in >98%, acylation with hindered carboxylic acids generally resulted in ca. 80% efficiencies.

8. Acidolytic treatment using DCM:hexafluoroisopropanol (1:1) for 2 h at ambient temperature afforded the hydroxamic acid in only 45% yield. However, it is worth noting that the tethered Fmoc-N(Pr)-*O*-2-chlorotrityl polystyrene, on treatment with similar acidolytic cocktail effected quantitative release of Fmoc-N(Pr)-OH.

9. An OMNI Fit (1.0×10.0 cm) reaction column was used. Alternatively, this can be carried out using either the Quest 210 semiautomated synthesizer or the Advanced ChemTech peptide synthesizer.

10. The purity of peptides obtained generally varies (50–90%) with the assembled peptide sequence. Owing to the protracted 90% TFA treatment, the major impurity usually observed is the acid-catalyzed decomposition product, peptidyl acid—the quantity of this undesired side product varies with peptide sequence and, particularly, the C-terminus amino acid residue.

DISCUSSION

Naturally occurring pseudopeptidyl hydroxamic acids e.g., actinonin, foroxymithine, propioxatins, and matlystatin B[5] and synthetic hydroxamic acids[6] are potent and selective inhibitors of many important metalloproteases, including matrix metalloproteases, angiotensin-converting enzyme, endothelin-converting enzyme, and enkephalinases. Inhibition of these proteases, which house a zinc atom within the catalytic domain, is the result of the ability of the hydroxamic acid functionality to form a bidentate chelate with the zinc atom. The sheer numbers of these endogenous metalloproteases, which are involved in a diverse range of biologic processes suggest that these enzymes are valuable targets for inhibition within the context of therapeutic intervention.

Hence, the implication of combinatorial chemistry for high throughput generation of structurally diverse hydroxamic acids is self-evident. Several solid-phase approaches for their syntheses have been reported,[1,7–11] the majority of which are based on the anchoring of *N*-hydroxyphthalimide onto an appropriate solid support. After hydrazine-mediated *N*-deprotection, *N*-acylation of the resin-bound hydroxylamine would yield the desired *O*-anchored hydroxamic acid, which is typically released by acidolysis.

In 1983, Prasad et al.[12] first reported the condensation of chloromethyl polystyrene with *N*-hydroxyphthalimide to give the ester, hydrazinolysis of which yielded the desired resin-bound hydroxylamine. However, the sole purpose of this reagent was to react with, and hence extract ketones from, a complex steroidal mixture, and its use for the solid-phase synthesis of hydroxamic acids was not explored. Recently, the exploitation of the above solid-phase approach for the synthesis of hydroxamic acids was independently reported by three groups,[7–9] all of which differ only in the method for the initial anchoring of *N*-hydroxyphthalimide to an 4-alkoxybenzyl alcohol functionalized polystyrene or trityl chloride polystyrene. Subsequent *N*-deprotection was achieved by prolonged treatment (12–18 h) with hydrazine hydrate in DMF to afford the key intermediate *O*-anchored hydroxylamine.

In contrast, we reported a facile and efficient method for the preparation of the key intermediate, aminooxy-2-chlorotrityl polystyrene, via the readily synthesized *N*-(9-fluorenylmethoxy-carbonyl)-hydroxyamine.[1] The compound Fmoc-NHOH was synthesized, in excellent yield as a white crystalline solid, by reacting hydroxylamine with stoichiometric amount of Fmoc-Cl under mild basic conditions for 3–4 h. Using the high loading 2-chlorotrityl chloride polystyrene,[2] Fmoc-NHOH was selectively *O*-anchored, via a simple S$_N$1 reaction, to afford the desired *N*-(9-fluorenylmethoxycarbonyl)aminooxy-2-chlorotrityl polystyrene. Typically, this condensation reaction was achieved in efficiency > 90%. Selective *O*-anchoring is achieved owing to the steric bulk of the trityl moiety. Conversely, it is worth noting that in our subsequent studies, condensation of Fmoc-NHOH with substituted benzhydryl chloride polystyrene gave a mixture of *O*- and *N*-anchored derivatives.

Moreover, during the course of our studies, *N*-[1-(4,4-di-methyl-2,6-dioxocyclohex-1-ylidene)ethyl]hydroxylamine, Dde-NHOH was also successfully coupled with 2-chlorotrityl chloride polystyrene in excellent efficiency.[1] The novel compound Dde-NHOH was prepared, in 51% yield, by reacting 2-acetyldi-medone with hydroxylamine in MeOH:THF at 5°C for 3 h, followed by recrystallization from ice-cold hexane; the major side-product, which increases in quantity over prolonged reaction time, was the predicted cyclized derivative 3,6,6-trimethyl-4-oxo-4,5,6,7-tetrahydro-1,2-benzisoxazole.

N-(9-Fluorenylmethoxycarbonyl)aminooxy-2-chlorotrityl polystyrene was then *N*-deprotected within minutes by treatment with 20% v/v piperidine in DMF to afford the key intermediate aminooxy-2-chlorotrityl polystyrene. With this in hand, *N*-acylation was then carried out and, where appropriate, followed by a series of chemical transformations to yield resin-bound hydroxamic acid derivatives; examples of these transformations were illustrated above.

In this synthetic strategy, release of the assembled resin-bound hydroxamic acid derivatives was efficiently achieved by exposure of the resin material to mild acidic reagents, including 1% v/v TFA in DCM for 10–15 min. Although we have had limited success, acidolytic release of the assembled molecule could also be effected by exposure to 50% v/v HFIP in DCM for 2 h. It is noteworthy that the use of mild acidolytic reagents in our solid-phase strategy is a significant advantage, because strong acidic reagents are known to cause decomposition of hydroxamic acids to the corresponding acids.

In conclusion, we anticipate that *N*-Fmoc-aminooxy-2-chlorotrityl polystyrene will prove an indispensable reagent for the solid-phase synthesis of hydroxamic acids by multiple and combinatorial approaches. Not only is its production both efficient and cost effective, but release of the assembled hydroxamic acid derivative is readily accomplished using mild acidolytic reagents.

REFERENCES

1. Mellor, S. L.; McGuire, C.; Chan, W. C. *Tetrahedron Lett.* **1997**, *38*, 3311.

2. Barlos, K.; Chatzi, O.; Gatos, D.; Stavropoulos, G. *Int. J. Peptide Protein Res.* **1991**, *37*, 513.

3. Carpino, L. A. *J. Am. Chem. Soc.* **1993**, *115*, 4397.

4. Carpino, L. A.; Sadat-Aalaee, D.; Chao, H. G.; DeSelms, R. H. *J. Am. Chem. Soc.* **1990**, *112*, 9651.

5. Umezawa, H.; Aoyagi, T.; Tanaka, T. et al. *J. Antibiotics* **1985**, *38*, 1629; Umezawa, H.; Aoyagi, T.; Ogawa, K. et al. *J. Antibiotics* **1985**, *38*, 1813; Inaoka, Y.; Takahashi, S.; Kinoshita, T. *J. Antibiotics* **1986**, *39*, 1378; and Tamaki, K.; Ogita, T.; Tanazawa, K.; Sugimura, Y. *Tetrahedron Lett.* **1993** *34*, 683.

6. Bihovsky, R.; Levison, B. L.; Loewi, R. C. et al. *J. Med. Chem.* **1995**, *38*, 2119 and Onishi, H. R.; Pelak, B. A.; Gerkens, L. S. et al. *Science* **1996**, *274*, 980.

7. Floyd, C. D.; Lewis, C. N.; Patel, S. R.; Whittaker, M. *Tetrahedron Lett.* **1996**, *37*, 8045.

8. Richter, L. S.; Desai, M. C. *Tetrahedron Lett.* **1997**, *38*, 321.

9. Bauer, U.; Ho, W.-B.; Koskinen, A. M. P. *Tetrahedron Lett.* **1997**, *38*, 7233.

10. Ngu, K.; Patel, D. V. *J. Org. Chem.* **1997**, *62*, 7088.

11. Mellor, S. L.; Chan, W. C. *Chem. Commun.* **1997**, 2005.

12. Prasad, V. V. K.; Warnes, P. A.; Lieberman, S. *J. Steroid Biochem.* **1983**, *18*, 257.

CHAPTER NINE

FACILE PREPARATION OF CHLOROMETHYLARYL SOLID SUPPORTS

Submitted by **DAVID A. NUGIEL, DEAN A. WACKER, and GREGORY A. NEMETH**

DuPont Pharmaceuticals, Box 80336 Wilmington,
DE, USA 19880-0336

Checked by JOACHIM DICKHAUT

Hoechst Schering AqrEvo GmbH, Hoechst Works (G-836), D-65926,
Frankfurt am Main, Germany

REACTION SCHEME

PROCEDURE

Wang resin (3.0 g, 0.9 mmol / g, 2.7 mmol; note 1) was suspended
in dry DMF (25 mL; note 2) to which diisopropylethylamine

(1.9 mL, 10.8 mmol; note 3) was added in one portion at room temperature. After 5 min, methanesulfonyl chloride (0.78 mL, 8.1 mmol; note 4) was added via syringe over 1 min. The addition causes an exothermic reaction. After 3 days, the resin was filtered and washed with DMF (2 × 20 mL), methanol (2 × 20 mL), and dichloromethane (2 × 20 mL). The resin was then dried in a vacuum oven at 60°C overnight. The amount of resin recovered was 2.95 g (note 5). Elemental analysis for chlorine: calculated, 3.19; observed, 3.27. Elemental analysis did not reveal any nitrogen, indicating that all the chlorine observed came from the resin. The IR spectrum showed no OH stretch, indicating complete disappearance of the benzylic alcohol. The ^{13}C NMR showed the complete disappearance of the hydroxymethyl benzylic carbon at 64.5 ppm with a new signal at 46.3 ppm corresponding to the newly formed chloromethyl benzylic substituent (note 6). The resin is stable at room temperature and can be stored indefinitely in a closed container.

NOTES

1. Purchased from NovaBiochem, Cat. # 01-64-0014.

2. Purchased from the Aldrich Chemical Company, Cat. # 22705-6.

3. Dried and distilled; purchased from the Aldrich Chemical Company, Cat. # 38764-9.

4. Purchased from the Aldrich Chemical Company, Cat. # 47125-9.

5. Shorter times typically led to incomplete conversion as shown in Table 9.1.

6. NMR taken in nondeuterated dichloromethane.

TABLE 9.1. Versatility of the Method

Resin	Reaction Time (h)	Conversion (%)	^{13}CNMR Shift (ppm)
Wang	72	100	46.3
SASRIN	72	100	48.1
Photocleavable AM[a]	24	90	44.3
Photocleavable AM	72	100	
Photocleavable TG[b]	72	100	43.7

[a] Hydroxymethyl-Photolinker AM resin.[5]
[b] Hydroxymethyl-Photolinker NovaSyn TG resin.[5]

DISCUSSION

There is a constant search for adapting different types of chemistry to solid supports. One approach to this goal is expanding the limited supply of commercially available solid supports. A previous report by Mergler et al.[1] disclosed a method for converting Wang[2] and SASRIN[3] resins to their corresponding chloromethylaryl analogs. This allowed loading amino acids onto the resin and subsequently coupling the amino acids with minimal racemization. Employing triphenylphosphine dichloride[4] to perform this conversion gave variable results and in only one case quantitative conversion to the desired chloromethylaryl resin. We disclose here a superior method of preparing chloromethylaryl resins, which consistently gives quantitative conversions.

Table 9.1 shows the method's versatility across several solid-support types. Care must be taken to dry the tentagel resins by lyophilization for 24 h before subjecting them to the reaction conditions. In the examples shown, quantitative conversions were obtained as determined by elemental analysis and ^{13}C NMR. The mild reaction conditions are most evident by the quantitative conversion of SASRIN resin to its corresponding chloromethyl derivative. The use of DMF was critical to the success of this procedure. The reaction did not proceed at all when dichloromethane or THF was employed. Stopping the reaction at less than 3 days showed incomplete conversion. This was not detrimental, because the resin could be resubjected to the reaction conditions, driving the reaction to completion.

REFERENCES

1. Mergler, M.; Nyfeler, R.; Gosteli, J. *Tetrahedron Lett.* **1989**, *30*, 6741, 6745.

2. Wang, S.-W. *J. Am. Chem. Soc.* **1973**, *95*, 1328.

3. Mergler, M.; Tanner, R.; Gosteli, J.; Grogg, R. *Tetrahedron Lett.* **1988**, *29*, 4005.

4. Appel, R.; *Angew. Chem.* **1975**, *87*, 863.

5. Holmes, C. P.; Jones, D. G. *J. Org. Chem.* **1995**, *60*, 2318.

CHAPTER TEN

PREPARATION OF AMEBA RESIN

Submitted by **PAUL C. FRITCH, ADAM M. FIVUSH, and TIMOTHY M. WILLSON**

Department of Medicinal Chemistry, Glaxo Wellcome Research and Development, P.O. Box 13398, Research Triangle Park, NC, USA 27709

Checked by LAXMINARAYAN BHAT and GUNDA I. GEORG

Medicinal Chemistry, School of Pharmacy, 4070 Malott Hall, Lawrence, KS, USA 66045-2506

REACTION SCHEME

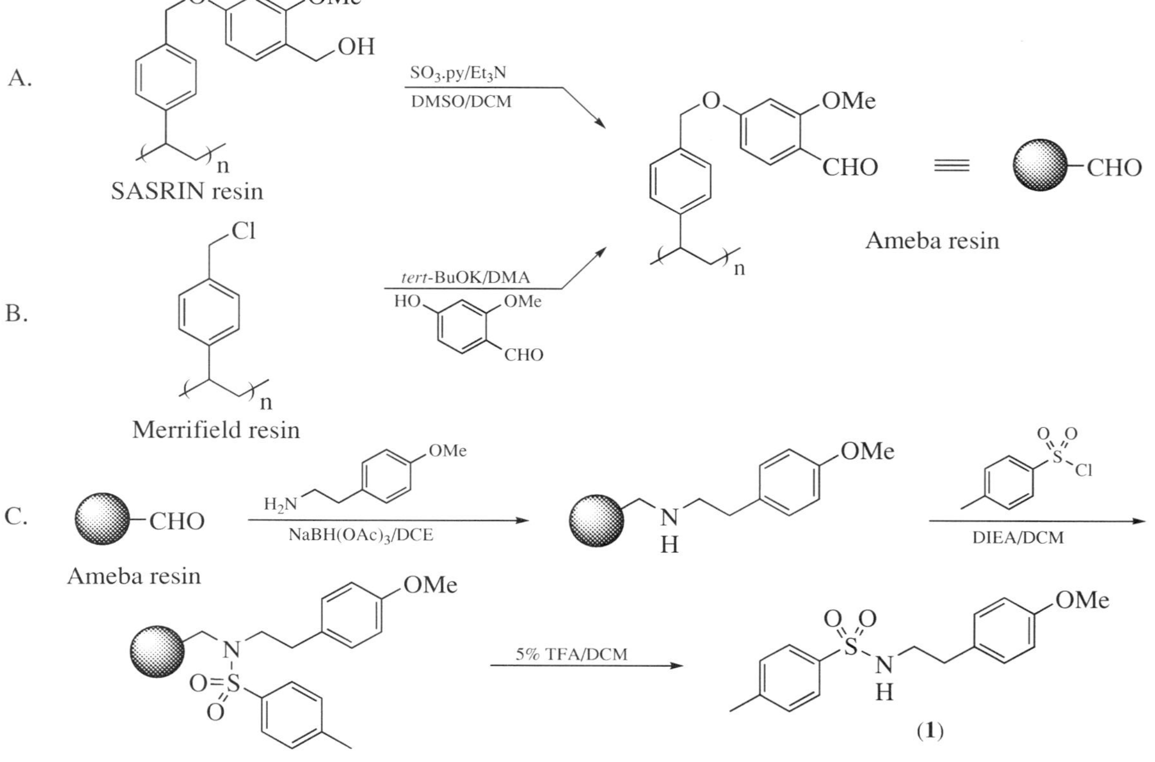

PROCEDURE

Preparation of AMEBA Resin A from SASRIN Resin, 200–400 Mesh[1]

A total of 10.0 g (8.9 mmol) SASRIN resin[2] (note 1) was washed with *N,N*-dimethylformamide (DMF; 2 × 25 mL), methanol (MeOH; 2 × 25 mL), and dichloromethane (DCM; note 2; 2 × 25 mL), and dried under vacuum (0.5 torr) at 70°C overnight. To a suspension of the dried SASRIN resin in 100 mL of methyl sulfoxide (DMSO; note 2) and 25 mL of DCM was added 12.4 mL (89 mmol, 10.0 Eq.) triethylamine (note 2) followed by 7.1 g (44.5 mmol, 5.0 Eq.) sulfur trioxide-pyridine complex (note 2). The suspension was shaken on a radial arm at room temperature overnight (note 3); filtered on a glass frit; and washed with DCM (3 × 100 mL), DMSO (3 × 100 mL), DCM (3 × 100 mL), and tetrahydrofuran (THF; 3 × 100 mL); and dried under vacuum (0.5 torr) at room temperature to give 10.0 g Ameba resin (notes 4 and 5).

Preparation of Ameba Resins Ba–Bd from Merrifield Resin[3]

A total of 1.00 g (0.57 mmol) Merrifield resin (LL, 100–200 mesh; note 6) was swollen in 5 mL *N,N*-dimethylacetamide (DMA; note 2) under N_2 for 20 min. A three-neck flask was charged under N_2 with 0.20 g (1.71 mmol, 3.0 Eq.) potassium *tert*-butoxide (note 2), 0.26 g (1.71 mmol, 3.0 Eq.) 4-hydroxy-2-methoxybenzaldehyde (note 2), and 5 mL DMA. The solution was stirred for 10 min and then added by syringe to the suspension of Merrifield resin. The reaction mixture was shaken and heated at 90°C for 4 h and 50°C overnight (note 7). The reaction mixture was cooled to room temperature; filtered on a glass frit; and washed with water (2 × 10 mL), methanol (MeOH; 2 × 10 mL), THF 2 × 10 mL), 2:1 water / THF (2 × 10 mL), water (2 × 10 mL), THF (2 × 10 mL), and MeOH (2 × 10 mL). Ameba resin **Ba** (Table 10.1) was dried under vacuum (0.5 torr) at room temperature overnight (notes 5 and 8).

Ameba resin **Bb** (Table 10.1) was prepared from 1.00 g (1.10 mmol) Merrifield resin (HL, 100–200 mesh; note 6) in 5.0 mL DMA and a solution of 0.38 g potassium *tert*-butoxide and 0.50 g 4-hydroxy-2-methoxybenzaldehyde in 9.5 mL of DMA.

Ameba resin **Bc** (Table 10.1) was prepared from 1.00 g (0.63 mmol) Merrifield resin (LL, 200–400 mesh; note 6) in 5.0 mL DMA and a solution of 0.22 g potassium *tert*-butoxide and 0.28 g 4-hydroxy-2-methoxybenzaldehyde in 5.5 mL of DMA.

Ameba resin **Bd** (Table 10.1) was prepared from 1.00 g (1.49 mmol) Merrifield resin (HL, 200–400 mesh; note 6) in 5.0 mL DMA and a solution of 0.52 g potassium *tert*-butoxide and 0.68 g 4-hydroxy-2-methoxybenzaldehyde in 13 mL DMA.

TABLE 10.1. Ameba Resin Loading Values and Yields for Sulfonamide (1)

Ameba Resin	Prepared From (mesh)	Starting Resin Loading (mmol/g)		Calculated Loading of AMEBA Resin (mmol/g)		Sulfonamide Yield (%)	
		Submitter	Checker	Submitter	Checker	Submitter	Checker
A	SASRIN (200–400)	0.89	1.02	0.89	1.02	66	69
Ba	Merrifield (LL 100–200)	0.57	0.57	0.53	0.53	85	85
Bb	Merrifield (HL 100–200)	1.10	1.48	0.98	1.26	93	67
Bc	Merrifield (LL 200–400)	0.63	0.63	0.59	0.59	65	91
Bd	Merrifield (HL 200–400)	1.49	1.24	1.27	1.08	81	70

Evaluation of Ameba Resins

For the preparation of *N*-[2-(methoxyphenyl)ethyl]-4-methylben-
zenesulfonamide (**1**) from Ameba resins **A** and **Ba–Bd**, 100 mg
(0.089 mmol) Ameba resin **A** was added to a glass peptide reaction
vessel, suspended in 3.0 mL 1,2-dichloroethane (DCE; note 2),
and treated with 26 µL (0.18 mmol, 2.0 Eq.) 2-(4-methoxy-
phenyl)ethylamine (note 2) and 38 mg (0.178 mmol, 2.0 Eq.)
sodium triacetoxyborohydride (note 2). The suspension was
shaken for 1 h; treated with 5 mL MeOH; filtered on a glass frit;
and washed with DCM (2 × 5 mL), DMF (2 × 5 mL), MeOH (2 ×
5 mL), and DCM (2 × 5 mL). The resin was dried under vacuum
(0.5 torr) at room temperature overnight. The resin was suspended
in 1.5 mL DCM, treated with 155 µL (0.89 mmol, 10.0 Eq.) *N,N*-
diisopropylethylamine (note 2) and 85 mg (0.445 mmol, 5.0 Eq.)
p-toluenesulfonyl chloride (note 2), and shaken for 3.5 h. The
reaction mixture was filtered on a glass frit, washed with DCM (2
× 5 mL), DMF (2 × 5 mL), MeOH (2 × 5 mL), and DCM (2 ×
5 mL), and dried under vacuum (0.5 torr) at room temperature for
2 h. The resin was treated with 2.5 mL of a solution of 5%
trifluoroacetic acid (note 2) in DCM, shaken for 15 min, filtered
on a glass frit, and washed with DCM (3 × 5 mL). The combined
filtrate and washings were concentrated and dried under vacuum
(0.5 torr) at room temperature overnight to afford 18.0 mg (66%)
N-[2-(methoxyphenyl)ethyl]-4-methylbenzenesulfonamide (**1**).

Using the procedure described above, 168 mg (0.089 mmol)
Ameba resin **Ba**, 91 mg (0.089 mmol) Ameba resin **Bb**, 151 mg
(0.089 mmol) Ameba resin **Bc**, and 70 mg (0.089 mmol) Ameba
resin **Bd** yielded 23.1 mg (85%), 25.5 mg (93%), 17.6 mg (65%),
and 21.9 mg (81%) of sulfonamide (**1**), respectively (Table 10.1).

NOTES

1. SASRIN resin (0.89 mmol/g) was obtained from Bachem
   Bioscience, Inc., Product # D-1295, Lot # 507127. Checkers

used SASRIN resin (1.02 mmol/g) obtained from Bachem Bioscience, Inc., Product # D-1295, Lot # 516349.

2. DCM (anhydrous), DMSO (anhydrous), triethylamine (99+%), sulfur trioxide-pyridine complex, DMA (anhydrous), potassium *tert*-butoxide (95%), DCE (anhydrous), 2-(4-methoxyphenyl)ethylamine (98+%), sodium triacetoxyborohydride (95%), *N,N*-diisopropylethylamine (99%), *p*-toluenesulfonyl chloride (99+%), and trifluoroacetic acid (99+%) were obtained from Aldrich Chemical Company, Inc. 4-Hydroxy-2-methoxybenzaldehyde (>98%) was obtained from Fluka Chemie, AG.

3. The checkers used a LabLine orbit shaker at 200 rpm.

4. Ameba resin **A** loading was assumed to be 0.89 mmol/g, based on the loading of the starting SASRIN resin.

5. Ameba resin was characterized by the diagnostic aldehyde signal at 10.5 ppm using Nanoprobe ^{1}H NMR.[4] Checkers characterized Ameba resin by the diagnostic aldehyde signals at 1675–1684 cm^{-1} using IR.

6. Merrifield resin was obtained from Novabiochem: LL (100–200 mesh), 0.57 mmol/g, Product # 01-64-0008, Lot # A18613; HL (100–200 mesh), 1.10 mmol/g, Product # 01-64-0070, Lot # A16109; LL (200–400 mesh), 0.63 mmol/g, Product # 01-640007, Lot # A18806; HL (200–400 mesh), 1.49 mmol/g, Product # 01-64-0002, Lot # A16226. Checkers obtained Merrifield resin from Novabiochem: LL (100–200 mesh), 0.57 mmol/g, Product # 01-64-0008, Lot # A18613; HL (100–200 mesh), 1.48 mmol/g, Product # 01-64-0070, Lot # A20333; LL (200–400 mesh), 0.63 mmol/g, Product # 01-64-0007, Lot No. A18806; HL (200–400 mesh), 1.24 mmol/g, Product # 01-64-0002, Lot # A17484.

7. A DIGI-BLOCK^{T} Jr. heating block (Laboratory Devices, USA, Inc.) that was fitted to a IKA-Schuttler-MTS-2 orbital

shaker was used. Checkers used a LabLine orbit shaker and a Thermolyne heating block.

8. Ameba resin **B** loading was calculated using the following formula: New loading = $(1/1 + (MW \times$ old loading$/1000)) \times$ old loading; where MW is the additional molecular weight of the compound added to the resin ($152 - 36.5 = 115.5$).

DISCUSSION

Ameba resin has been employed for the solid-phase organic synthesis of amides, sulfonamides, ureas, and carbamates by reductive amination and subsequent *N*-derivatization. The resin is acid sensitive, so that the products can be cleaved under mild conditions with dilute solutions of trifluoroacetic acid.[2,5,6] The procedures described above illustrate two methods for the preparation of Ameba resin. The efficiency of the prepared resins was evaluated by comparing the yield of *N*-[2-(4-methoxyphenyl)-ethyl]-4-methylbenzenesulfonamide, which was synthesized on the resins. Procedure A employed the oxidation of commercially available SASRIN resin with sulfur trioxide-pyridine complex.[1] Ameba resin **A** synthesized by this method afforded a 66% yield of the sulfonamide (**1**), indicating either incomplete oxidation in the preparation of the resin or incomplete reaction in the synthesis of the sulfonamide. The cost of preparing the resin by procedure A is estimated at $52/mmol, with the major expense being the cost of SASRIN resin. Procedure B, which was based on the report of Katritzky et al.,[3] employed the coupling of commercially available 4-hydroxy-2-methoxybenzaldehyde with four Merrifield resins of different mesh size and loading. Ameba resins **Ba–Bd** synthesized by this method were also evaluated by preparation of sulfonamide (**1**). We found that the 100–200 mesh resins afforded slightly superior yields of the product compared to the 200–400 mesh resins (Table 10.1). The checkers found that both mesh sizes of the LL resins gave slightly higher yields than the HL resins (Table 10.1). Thus procedure B is generally applicable to all four forms of Merrifield resin. Ameba resins **Ba–Bd** can be prepared for $2–4/mmol, depending on the loading of the Merrifield resin employed. Although Ameba resin is commercially available from Fluka Chemie AG (Product # 09942) for $22/mmol, preparation by procedure B represents a cost-effective source of this acid-sensitive aldehyde resin.

REFERENCES

1. Fivush, A. M.; Willson, T. M. *Tetrahedron Lett.* **1997**, *38*, 7151.

2. Mergler, M.; Nyfeler, R.; Gostelli, J.; Grogg, P. *Chem. Biol., Proc. Am. Pept. Symp. 10th* **1988**, 259.

3. Katritzky, A. R.; Toader, D.; Watson, K.; Kiely, J. S. *Tetrahedron Lett.* **1997**, *38*, 7849.

4. Keifer, P. A. *J. Org. Chem.* **1996**, *61*, 1558.

5. Kiselyov, A. S.; Smith, L.; Virgilio, A.; Armstrong, R. W. *Tetrahedron* **1998**, *54*, 7987.

6. Ouyang, X.; Tamayo, N.; Kiselyov, A. S. *Tetrahedron* **1999**, *55*, 2827.

CHAPTER ELEVEN

AN EFFICIENT SOLID-PHASE SYNTHETIC ROUTE TO 1,3-DISUBSTITUTED 2,4(1H,3H)-QUINAZOLINEDIONES SUITABLE FOR COMBINATORIAL SYNTHESIS

Submitted by **ADRIAN L. SMITH and JOSEPH G. NEDUVELIL**

Merck Sharp & Dohme Research Laboratories, Neuroscience Research Centre, Terlings Park, Harlow, Essex CM20 2QR, United Kingdom

Checked by SHARON A. JACKSON, DONGLIANG ZHAN, and TASIR S. HAQUE

The DuPont Pharmaceuticals Company, Department of Chemical & Physical Sciences, Experimental Station, P.O. Box 80500, Wilmington, DE, USA 19880-0500

LIBRARY SYNTHESIS ROUTE

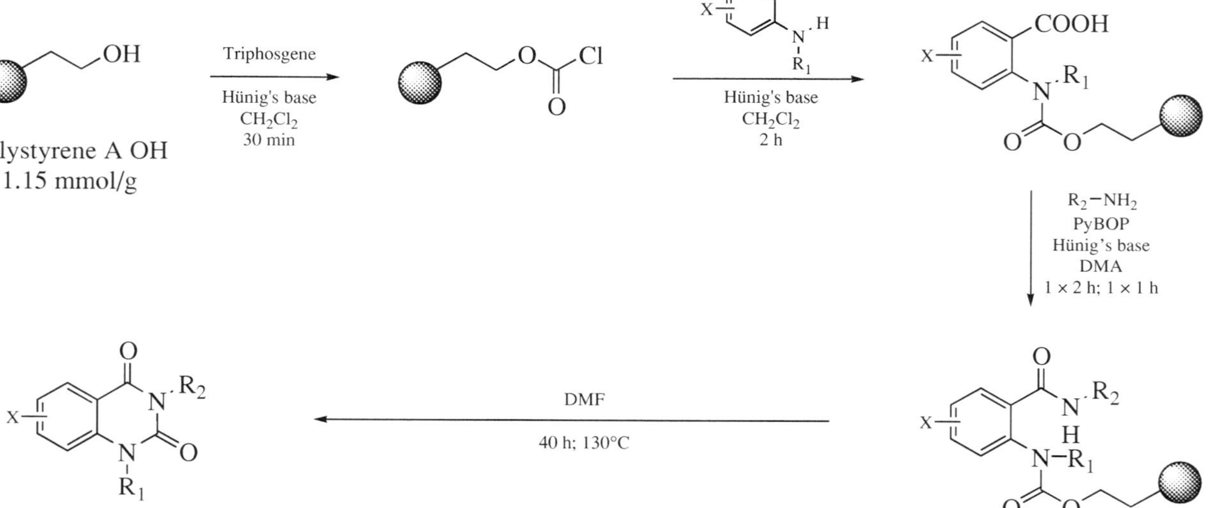

BUILDING BLOCKS

Anthranilic acids:

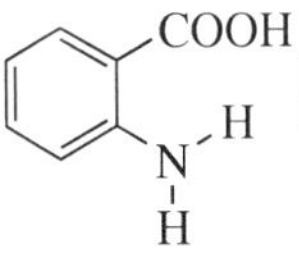

(1) 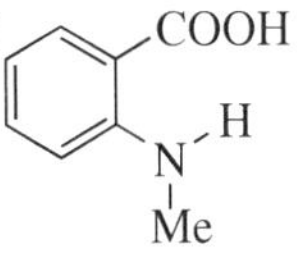   (2) 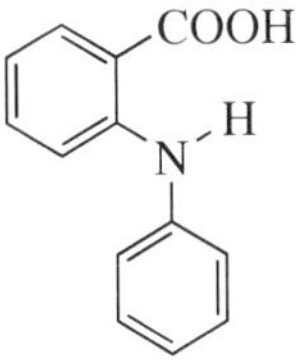   (3) 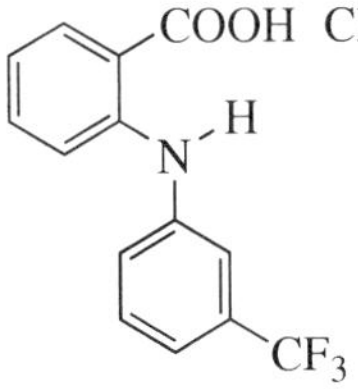   (4) 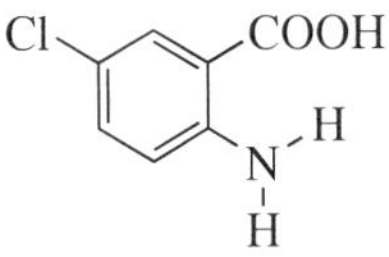   (5)

Amines R_2-NH_2:

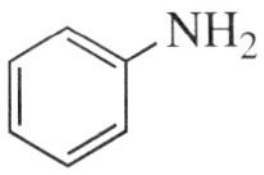

(6) 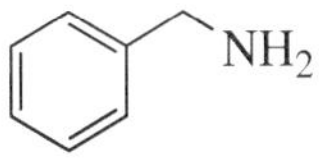   (7)    (8)    (9) 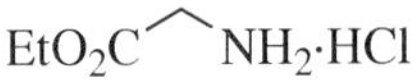   (10)

PROCEDURE

1. Polystyrene A OH (3.0 g; 1.15 mmol/g) was suspended in DMF/CH$_2$Cl$_2$ (1:1, 30 mL total volume) and 1.00 mL aliquots(100 mg resin; 0.115 mmol) were added by Gilson pipette (note 1) to 25 individual Quest 210 reactor vessels (5 mL volume; note 2). The reactors were washed with CH$_2$Cl$_2$ (10 × 2 mL), using dry nitrogen gas from the Quest manifold to drain the reactors between washings. The resulting resin was suspended in CH$_2$Cl$_2$ (0.9 mL) and 1.00 mL of a stock solution of triphosgene in CH$_2$Cl$_2$ (2.05 g, 6.9 mmol in 30 mL total volume; 0.23 mmol/reaction) was added to each reactor by Gilson pipette, followed by 87 μL (0.50 mmol) *N,N*-diisopropylethylamine (Hünig's base). The resins were mixed at ambient temperature (23°C) for 30 min, drained, and washed with CH$_2$Cl$_2$ (5 × 2 mL; note 3). A few resin beads were sampled from the first reactor and analyzed by diffuse reflectance FT-IR (note 4) to confirm complete reaction.

2. Stock solutions of the anthranilic acids **1** to **5** (note 5) in CH_2Cl_2 were prepared by dissolving 2.4 mmol of each anthranilic acid in 10 mL CH_2Cl_2 and 1 mL Hünig's base. A total of 2.00 mL of the appropriate anthranilic acid solution was added to each reactor (0.4 mmol). The resins were mixed at ambient temperature for 2 h; drained; and washed with CH_2Cl_2 (5 × 2 mL), MeOH (5 × 2 mL), and DMA (10 × 2 mL) (note 6). A few resin beads were sampled for each anthranilic acid used and analyzed by diffuse reflectance FT-IR to confirm complete reaction.

3. Stock solutions of amines **6–10** in DMA were prepared by dissolving 4.14 mmol of each amine in 11 mL DMA and 1 mL Hünig's base (10 mL DMA / 2 mL Hünig's base were used for the hydrochloride salt **10**). Aliquots, 1.00 mL (0.345 mmol), of the appropriate amine solution were added to each reactor followed by 1.00 mL 0.345 M PyBOP in DMA (0.345 mmol). The resins were mixed for 2 h and drained; then equal amounts of amine and PyBOP solutions were added and the resins were mixed for another 1 h (note 7). The reactions were drained and washed with DMA (5 × 2 mL), MeOH (5 × 2 mL), and DMF (10 × 2 mL).

4. The resulting DMF-swollen resins were heated at 130°C for 40 h (note 8) and allowed to cool to 80°C; the products were collected into test tubes by washing with DMF at 80°C (4 × 0.5 mL) (Note 9), allowing the resin to stand for 5 min between each addition of DMF and collection of washings. The resulting DMF product solutions were concentrated in vacuo (notes 10 and 11) to give off-white/brown products (highly crystalline in most cases).

Description of Solid-Phase Support

Polystyrene A OH: Loading 1.15 mmol/g; 200 – 400 mesh.

Rapp polymere Cat. # HA 1 400 00; Batch # 400s69.

NOTES

1. It is found to be advantageous to cut the bottom 2–3 mm from a 1-mL Gilson pipette tip with scissors when transferring resin slurries, otherwise blockage of the tip occurs with the swollen resin.

2. Available from Argonaut Technologies. Reagent additions were carried out through the luer ports on the upper manifold to maintain an anhydrous atmosphere during the reaction sequence. The chemistry described is very robust and can be carried out in any suitable solid phase reactor.

3. The chloroformate resin is readily prepared immediately before use and hence its stability to long-term storage has not been explored. However, no special precautions were needed in handling the resin and no stability problems were observed during the course of this work. The FT-IR spectrum of sampled beads showed no sign of a hydroxyl signal.

4. A Perkin Elmer Diffuse Reflectance Accessory Cat. # L127-5000 was used on a Perkin Elmer Spectrum 1000 FT-IR.

5. *N*-Methyl anthranilic acid (**2**) as supplied by Aldrich (Cat. # 13,706-5) contains 5% anthranilic acid (**1**) and must be purified by recrystallization from EtOH before use. The remaining anthranilic acids were used as received from Aldrich.

6. Anhydrous DMA was used as supplied by Aldrich (Cat. # 27,101-2).

7. The coupling times given are generally found to be sufficient, although longer reaction times may be beneficial for unreactive systems.

8. The swollen resin contains $\sim$0.5 mL DMF, which is sufficient for the reaction. Agitation is not necessary. There is some scope for changing the precise cleavage conditions; for

most reactions it is found that the limiting temperature at which thermal cyclization / cleavage occurs is approximately 100°C, with synthetically useful yields (generally 40–60%) being obtained at 125°C for 16 h. It is found that products are generally thermally robust with a wide range of substituents, and in these cases cleavage can be carried out at 150°C for 2–4 h. The conditions described in this procedure are designed to give near-maximum yields.

9. Quinazolinediones tend to be highly crystalline and consequently can be difficult to dissolve.

10. A Savant AES2000 SpeedVac was used on high setting for 2 h.

11. Although the checkers reported that the procedure gave exceptionally pure crude products in a reliable and reproducible manner, they observed substantially reduced yields of both crude and purified products (reported purified yields were variable and typically in the range of 10–40%). One possible reason for this is the higher substitution level of the resin used (1.45 mmol/g), which the submitters feel is probably too high for carrying out the chemistry efficiently. During the development of this chemistry, an alternative chloroformate resin was originally used based on functionalization of aminomethyl polystyrene resin with tri(ethylene glycol) bis(chloroformate).[1] The original batch of aminomethyl polystyrene used (loading 0.6 mmol/g) gave reasonable yields of products; however, a second batch of resin was received with a much higher loading (1.2 mmol/g), resulting in very poor yields (<10%), primarily owing to extensive cross-linking.

DISCUSSION

An experimentally simple synthesis of 2,4(1*H*,3*H*)-quinazolinediones is described[1] that uses a thermal cyclization/cleavage as

TABLE 11.1.   Purity and Yield Data

Reaction	Anthranilic Acid	Amine	Crude Product Purity (%)	Purified Product Yield (%)	CAS Registry Number
1	1	6	57	11	150111-45-8
2	1	7	90	41	1932-42-9
3	1	8	91	44	10341-86-3
4	1	9	92	53	20297-19-2
5	1	10	90	49	58004-83-4
6	2	6	91	60	1028-37-1
7	2	7	98	78	199587-91-2
8	2	8	94	75	
9	2	9	99	81	
10	2	10	99	93	110679-30-6
11	3	6	93	64	89267-53-8
12	3	7	96	76	84-587-31-5
13	3	8	96	66	
14	3	9	96	39	
15	3	10	97	65	34928-91-1
16	4	6	89	49	56345-63-2
17	4	7	98	62	34929-05-0
18	4	8	92	62	34934-20-8
19	4	9	99	48	34934-15-1
20	4	10	77	47	39030-93-8
21	5	6	81	36	13191-02-1
22	5	7	90	50	209604-28-4
23	5	8	92	62	209604-17-1
24	5	9	88	50	209604-19-3
25	5	10	96	91	136148-77-7

the final step, resulting in a traceless solid phase synthesis.[2] In principle, only the desired products should be obtained during the cleavage step, because any incomplete reaction in the preceding reaction steps should give material incapable of cleaving from the resin. In practice, assuming efficient coupling reactions, the purity of the products depends on how well the cyclization proceeds. This, in turn, depends on how efficiently the cleavage transition state is achieved, being more efficient when R_1 is bulky (i.e. *not* H) and R_2 is not. The main limitation for the synthetic protocol as described here is with sterically encumbered amines R_2-NH_2; secondary alkyl amines such as cyclohexylamine give very poor reaction. Arylamines such as aniline generally work, although they can be borderline when combined with R_1 = H (cf. reactions 1 and 21 in Table 10.1). Otherwise, most amines tested work (if they are sufficiently nucleophilic to couple with the anthranilic acid). Virtually all anthranilic acids tested work efficiently. Yields of pure products isolated by preparative HPLC together with HPLC purities (diode array detector: 210–250 nm) of crude products are given in Table 11.1.

REFERENCES

1. Smith, A. L.; Thomson, C. G.; Leeson, P. D. *Bioorg. Med. Chem. Lett.* **1996**, *6*, 1483 and Smith, A. L. US Pat. 5,783,698 (July 21, 1998).

2. Camps, E.; Cartells, J.; Pi, J. *Anales de Quimica* **1974**, *70*, 848.; DeWitt, S. H.; Kiely, J. S.; Stankovic, C. J. et al. *Proc. Natl. Acad. Sci. USA* **1993**, *90*, 6909; Gordon, D. W.; Steele, J. *Bioorg. Med. Chem. Lett.* **1995**, *5*, 47; and Dressman, B. A.; Spangle, L. A.; Kaldor, S. W. *Tetrahedron Lett.* **1996**, *37*, 937.

CHAPTER TWELVE

BACKBONE AMIDE LINKER (BAL) STRATEGY FOR SOLID-PHASE SYNTHESIS

Submitted by JORDI ALSINA,[†] KNUD J. JENSEN, MICHAEL F. SONGSTER, JOSEF VÁGNER, FERNANDO ALBERICIO, and GEORGE BARANY[‡]

[†]Department of Organic Chemistry, University of Barcelona, 08028 Barcelona, Spain
[‡]Department of Chemistry, University of Minnesota, Minneapolis, Minnesota, USA 55455

Checked by JOHN FLYGARE and MONICA FERNANDEZ

Genentech, 1 DNA Way, South San Francisco, CA, USA 94080

REACTION SCHEMES

Preparation of *p*-PALdehyde [5-(4-formyl-3,5-dimethoxyphenoxy)-valeric acid], or *o,p*-PALdehyde [5-(4 or 2)-formyl-3,5-dimethoxy-phenoxy)valeric acid]

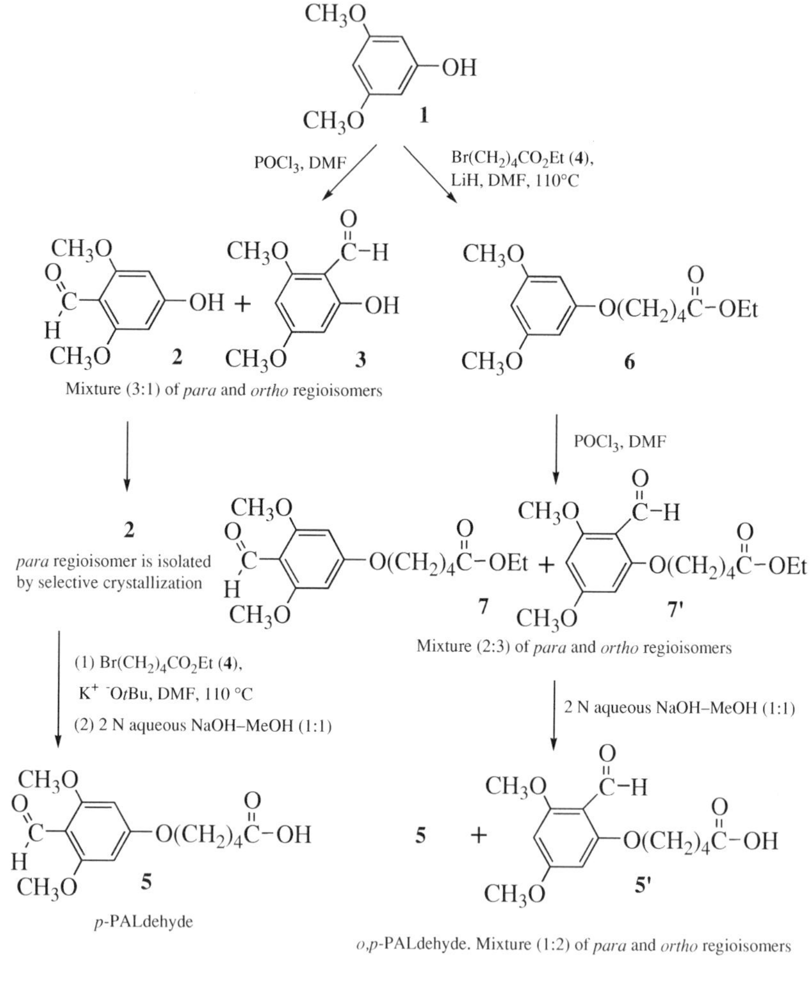

Scheme A

Preparation of BAL-anchored peptide-resins by on-resin reductive amination followed by stepwise chain elongation

p-PALdehyde (**5**)   + H_2N-®

BOP/HOBt/DIEA (2:3:3) in DMF

p-PALdehyde resin

(e.g., H-Phe-OtBu-HCl and H-Ala-OAllyl-HCl)

+

$NaBH_3CN$ in DMF

$R^1 = CO_2tBu$ or CO_2Allyl; $R^2 = CH_2$-Ph or CH_3

Prot = Fmoc if $R^1 = CO_2tBu$
Prot = Trt if $R^1 = CO_2Allyl$

Fmoc-Leu-OH/HATU/DIEA (10:10:20) or Trt-Gly-OH/PyAOP/DIEA (10:10:20) in CH_2Cl_2–DMF (9:1)

R^2 and R^3 = appropriate side-chain functionality

Prot = Trt
$R^1 = CO_2Allyl$

TFA–H_2O–CH_2Cl_2 (2:1:97), 5 min.

$R^2 = CH_3$ and $R^3 = H$.

$R^1 = CO_2Allyl$

Fmoc-Glu(OtBu)-OH/PyAOP/DIEA (10:10:20) in DMF

$R^2 = CH_3$ and $R^3 = H$.

$R^1 = CO_2Allyl$

Scheme B

PROCEDURE

PREPARATION OF *p*-PALDEHYDE [5-(4-FORMYL-3,5-DIMETHOXYPHENOXY)VALERIC ACID] (SCHEME A)

4-Formyl-3,5-dimethoxyphenol (2) (Pure Isomer) (Note 1)

The viscous mixture formed from 3,5-dimethoxyphenol (**1**) (20 g, 0.13 mol) and phosphorous oxychloride (24.2 mL, 0.26 mol) is stirred mechanically (Note 2) at 0°C, and *N,N*-dimethylformamide (DMF) (15 mL, 0.2 mol) is added portionwise over 0.5 h. The reaction mixture is stirred for an additional 15 h at 25°C and then quenched by pouring over ice (300 g). The very acidic aqueous solution is washed with ethyl ether (3 × 200 mL), and the aqueous phase is filtered to remove a tan residue (2.7 g, 11%), which by NMR (CD$_3$SOCD$_3$) is mainly 2-formyl-3,5-dimethoxyphenol (**3**); ^{1}H NMR (CD$_3$SOCD$_3$) δ 10.02 (s, 1 H), 6.16 (s, 1 H), 6.11 (s, 1 H), 3.87 (s, 3 H), 3.85 (s, 3 H); ^{13}C NMR (CD$_3$SOCD$_3$) δ 190.9 (formyl), 167.7, 164.7, and 163.2 (aryl C1, C3, C5, not further assigned), 104.9 (aryl C2), 92.7 (aryl C4), 90.2 (aryl C6), 55.6 (CH$_3$O), 55.5 (CH$_3$O), admixed with some 2,6-diformyl-3,5-dimethoxyphenol. The filtrate is diluted with water (250 mL), and the pH is adjusted to 6.0 with 19 N aqueous NaOH (53 mL). A heavy precipitate forms, which is collected after 15 min on a Büchner funnel, washed with warm (32°C) ethyl ether (4 × 100 mL) to extract away NMR-pure **3** (1.9 g after concentration, 8%), and dried in vacuo. Yield: 13.1 g (52%) of title product, a whitish-tan powder (~95% pure), which is dissolved in hot ethanol (250 mL) to provide, after cooling, an 85% recovery of NMR-pure (>99%) title product; melting point 224–226°C [literature melting point 222–224°C]; ^{1}H NMR (CD$_3$SOCD$_3$) δ 10.16 (s, 1 H), 6.09 (s, 2 H), 3.76 (s, 6 H); ^{13}C NMR (CD$_3$SOCD$_3$) δ 184.8 (formyl), 164.8 (aryl C1), 163.2 (aryl C3 and C5), 106.4

(aryl C4), 91.4 (aryl C2 and C6), 55.2 (CH_3O). Analysis calculated for $C_9H_{10}O_4$; MW 182.18: C, 59.34; H, 5.53. Observed: C, 59.17; H, 5.57. The title procedure for **2** is readily scaled up 10-fold, with similar yields and purities.

5-(4-Formyl-3,5-dimethoxyphenoxy)valeric acid (5) (*p*-PALdehyde)

Method A

A mixture of 4-formyl-3,5-dimethoxyphenol (**2**) (3.65 g, 20 mmol) and potassium *tert*-butoxide (2.25 g, 20 mmol) in toluene (20 mL) is refluxed for 5 h under magnetic stirring. The toluene is removed by rotary evaporation, and ethyl 5-bromovalerate (**4**) (4.8 mL, 30 mmol) and DMF (50 mL) are added (note 3). The reaction mixture is stirred magnetically for 15 h at 110°C, after which the solvent is removed at 60°C (1 mm) to provide an oil (9.8 g), which lacks starting phenol but contains excess bromovalerate as well as the ester precursor. This entire oil is dissolved in 2 N aqueous NaOH–methanol (1:1, v/v) (130 mL). The solution is stirred for 30 min at 25°C and then diluted with EtOAc (total 200 mL) and water (200 mL), and the organic phase is discarded. The aqueous phase is acidified with 12 N aqueous HCl to pH 1 and extracted with EtOAc (1 × 200 mL + 2 × 100 mL). The combined organic phases are washed with saturated aqueous NaCl (2 × 100 mL), dried ($MgSO_4$), and concentrated to give an orange powder (4.76 g, 85%). An analytical sample is obtained by crystallization from hot acetone, hexane added at 25°C for first crop, and further chilling to 4°C for second crop. This gives a pale yellow solid (overall 80% recovery): melting point 130–132°C; ^{1}H NMR (CD_3SOCD_3) δ 10.20 (s, 1 H), 6.26 (s, 2 H), 4.1 (broad t, 2 H), 3.82 (s, 6 H), 2.3 (broad t, 2 H), 1.6–1.8 (m, 4 H). Analysis calculated for $C_{14}H_{18}O_6$, MW 282.29: C, 59.57; H, 6.43. Observed: C, 59.62; H, 6.36.

Method B

A mixture of 4-formyl-3,5-dimethoxyphenol (**2**) (3.48 g, 19.0 mmol), K_2CO_3 (3.94 g, 28.5 mmol), and ethyl 5-bromovalerate (**4**) (5.96 g, 28.5 mmol) is refluxed in 3-methyl-2-butanone (20 mL; boiling point 95°C) for 21 h, filtered at 25°C, and concentrated at 40°C (2 mm). The resultant golden-brown oil (6.27 g), which includes excess **4** but only trace **2**, is dissolved in methanol (32 mL), and 2 N aqueous NaOH (32 mL) is added. The solution is stirred for 30 min, diluted with water (60 mL), partially concentrated at 30°C (12 mm), and extracted with EtOAc (3 × 30 mL). The aqueous phase is brought to pH 2 with 12 N aqueous HCl (4.2 mL) and extracted with EtOAc (3 × 40 mL). The organic extracts are dried ($MgSO_4$) and concentrated to provide a semi-solid (3.35 g, 69%). NMR (CD_3SOCD_3) as before.

5-(2-Formyl-3,5-dimethoxyphenoxy)valeric acid (5′) (*o*-PALdehyde)

A mixture of 2-formyl-3,5-dimethoxyphenol (**3**) (8.0 g, 44 mmol), K_2CO_3 (9.12 g, 66 mmol), and ethyl 5-bromovalerate (**4**) (13.8 g, 66 mmol) is reacted and worked up following method B for **5**. The initial semisolid product (11.9 g, 96%) is dissolved in hot EtOAc (85 mL), and hexane (75 mL) is added portionwise to incipient turbidity. Crystals formed at 25°C are collected after 12 h: yield 6.7 g (55% overall for two steps); melting point 103–104°C; ^1H NMR (CD_3SOCD_3) δ 10.23 (s, 1 H), 6.26 (s, 1 H), 6.25 (s, 1 H), 4.05 (t, J = 5.9 Hz, 2 H), 3.86 (s, 3 H), 3.81 (s, 3 H), 2.29 (t, J = 7.1 Hz, 2 H), 1.6–1.8 (m, 4 H); ^{13}C NMR (CD_3SOCD_3) δ 185.7 (formyl), 174.3 (COOH), 165.9 (aryl C1), 163.1 (aryl C3 and C5), 108.1 (aryl C2), 91.3 and 90.8 (aryl C4 and C6), 68.1 (OCH_2), 55.9 and 55.7 (2 CH_3O), 33.2 (CH_2 α to COOH), 27.9 and 21.2 (valeryl side chain). Analysis calculated for $C_{14}H_{18}O_6$, MW 282.28: C, 59.56; H, 6.43. Observed: C, 59.71; H, 6.32.

Ethyl 5-(3,5-dimethoxyphenoxy)valerate (6)

A mixture of 3,5-dimethoxyphenol (**1**) (20 g, 0.13 mol), ethyl 5-bromovalerate (**4**) (27.2 g, 0.13 mol), and lithium hydride (1.56 g, 0.195 mol) in DMF (150 mL) is magnetically stirred overnight at 110°C. The solvent is then removed at 40°C and 2 mm, and the residual oil is taken up in EtOAc (100 mL). This is washed with saturated aqueous NaCl (3 × 40 mL), 2 N aqueous NaOH (2 × 40 mL), and saturated aqueous NaCl (3 × 40 mL); dried ($MgSO_4$); and evaporated to give an oil (19.8 g), which by NMR (CD_3SOCD_3) is a mixture of **6** and **4**. Unreacted phenol **1** is contained in the aqueous NaOH washings. The product mixture as obtained is used without further purification for the subsequent reaction. The ^{1}H NMR (CD_3SOCD_3) attributable to **6** δ 6.09 (s, 3 H), 4.06 (q, J = 7.1 Hz, 2 H), 3.93 (t, J = 5.7 Hz, 2 H), 3.72 (s, 6 H), 2.33 (t, J = 5.9 Hz, 2 H), 1.6–1.8 (m, 4 H), 1.19 (t, J = 7.1 Hz, 3 H). Compare with NMR (CD_3SOCD_3) of starting **1**: δ 5.97 (s, 3 H), 3.67 (s, 6 H) and of ethyl 5-bromovalerate δ 4.06 (q, J = 7.1 Hz, 2 H), 3.54 (t, J = 6.4 Hz, 2 H), 2.34 (t, J = 7.3 Hz, 2 H), 1.7–1.9 (m, 2 H), 1.5–1.7 (m, 2 H), 1.19 (t, J = 7.1 Hz, 3 H).

Ethyl 5-[(2 or 4)-formyl-3,5-dimethoxyphenoxy]valerate (7 and 7′)

The entire product from the previous reaction (calculated to contain about 46 mmol of **6**) is combined with phosphorus oxychloride (8.53 mL, 91.6 mmol). The viscous mixture is mechanically stirred (note 2) at 0°C, and DMF (5.31 mL, 68.7 mmol) is added portionwise over 1 h. The reaction mixture is stirred for an additional 20 h at 25°C, and then quenched by addition of ice (200 g). The very acidic aqueous solution is washed with ethyl ether (3 × 75 mL) to remove **4** carried over from the previous reaction, after which the pH is adjusted to 6.0 with 19 N aqueous NaOH. Sodium acetate (40 g) is also added, and the solution is extracted with EtOAc (3 × 75 mL). The combined organic phases are washed with saturated aqueous

NaCl (3 × 40 mL), dried (MgSO$_4$), and evaporated to give an oil (14.1 g), which is pure by NMR (CD$_3$SOCD$_3$) (2-formyl and 4-formyl isomers in 3:2 ratio); ^{1}H NMR (CD$_3$SOCD$_3$) δ 10.23 (s) and 10.21 (s) (major and minor isomer, respectively, total 1 H), 6.25 (apparent s, 2 H), 4.0–4.1 (m, 4 H), 3.86, 3.82 and 3.81 (three adjacent singlets, total 6 H), 2.3–2.5 (m, 2 H), 1.7–1.9 (m, 4 H), 1.19 and 1.18 (minor and major isomer, respectively, overlapping triplets, J = 7.1 Hz, 3 H).

5-[(2 or 4)-Formyl-3,5-dimethoxyphenoxy]valeric acid (5 and 5′)

Compounds **7** and **7′** (14 g of the pure oil, ca. 45 mmol) are dissolved in 2 N aqueous NaOH–methanol (1:1) (180 mL). After stirring for 30 min at 25°C, the solution is washed with EtOAc (3 × 75 mL) to remove some organic impurities, acidified with 12 N aqueous HCl to pH 2, and extracted with EtOAc (3 × 75 mL). The combined organic phases are washed with saturated aqueous NaCl (2 × 50 mL), dried (MgSO$_4$), and rotary evaporated to give an oil (10.5 g). An analytical sample is obtained by crystallization from hot EtOAc, pentane added at 25°C, and further chilling to 4°C. This gives a white solid, melting point 98–100°C; ^{1}H NMR (CD$_3$SOCD$_3$) δ 10.23 (s) and 10.20 (s) (major and minor isomer, respectively, ratio 2:1, total 1 H), 6.26 (s) and 6.25 (s) (total 2 H), 4.0–4.2 (m, 2 H), 3.87 (s), 3.83 (s) and 3.82 (s) (total 6 H), 2.2–2.3 (m, 2 H), 1.6–1.8 (m, 4 H). Analysis calculated for C$_{14}$H$_{18}$O$_6$, MW 282.29: C, 59.57; H, 6.43. Observed: C, 59.60; H, 6.49.

PREPARATION OF BAL-ANCHORED PEPTIDE RESINS BY ON-RESIN REDUCTIVE AMINATION, FOLLOWED BY STEPWISE CHAIN ELONGATION (SCHEME B)

Quantitative Coupling of *p*-PALdehyde or *o,p*-PALdehyde to an Amino-Functionalized Solid Support

Fmoc-Ile-PEG-PS resin (notes 4–6) (2.0 g, 0.24 mmol/g) is washed with DMF (2 × 2 min) and CH$_2$Cl$_2$ (2 × 2 min), and then treated with piperidine–DMF (1:4, 2 × 2 min, 1 × 15 min), followed by washings with DMF (5 × 2 min) and CH_2Cl_2 (2 × 2 min). Solid *o,p*-PALdehyde (0.26 g, 2 Eq.), benzotriazol-1-yl-*N*-oxy-tris(dimethylamino)phosphonium hexafluorophosphate (BOP; 0.43 g; 2 Eq.; note 7), and HOBt (0.19 g, 3 Eq.) are combined and dissolved in DMF (5 mL), DIEA (0.25 mL, 3 Eq.) is added, and after a 5-min preactivation, this solution is added to the resin. Coupling is allowed to proceed for 15 h, at which time the resin is only slightly positive to the Kaiser ninhydrin test.[1] The resultant *o,p*-PALdehyde-Ile-PEG-PS resin is washed with DMF (2 × 2 min) and CH_2Cl_2 (2 × 2 min), and then treated with acetic anhydride–DMF (1:9, 20 min), washed with DMF (5 × 2 min), CH_2Cl_2 (2 × 2 min), and MeOH (2 × 2 min), and finally dried in *vacuo*; aliquots are taken to test reductive amination as described immediately below.

Attachment of the C-Terminal Residue Through its Amino Group Via On-Resin Reductive Amination

The C^α-carboxyl group of an α-amino acid is suitably protected as required or alternatively, an amino-containing derivative with appropriate modification is used.

Method A

This method is used when the amino compound is a free amine (e.g., phenylalaninol *tert*-butyl ether, H-Phe-*ot*Bu). H-Phe-*ot*Bu (39 mg, 10 Eq.) and $NaBH_3CN$ (12 mg, 10 Eq., notes 8 and 9), dissolved together in HOAc–DMF (1:99, 0.5 mL), are added to the *o,p*-PALdehyde-Ile-PEG-PS resin (100 mg, 0.19 mmol/g) and reacted at 25°C for 18 h to give the H-(BAL-Ile-PEG-PS)Phe-*ot*Bu resin, which is washed consecutively with DMF (5 × 0.5 min), CH_2Cl_2 (3 × 0.5 min), DMF (3 × 0.5 min), piperidine–DMF (1:4, 3 × 1 min), DMF (5 × 0.5 min). and CH_2Cl_2 (3 × 0.5 min). It is then dried *in vacuo* and used as a starting point for manual chain assembly of peptides by protecting further protected amino acids. For calculating the yield (>95%) Phe-oh is not determined directly; rather the secondary amine is acylated by Fmoc-Gly-OH mediated with PyAOP/DIEA in DMF. Yields are calculated by amino acid analysis (note 10).

Method B

This method is used when the amino compound is a hydrochloride salt (e.g., H-Phe-O*t*Bu·HCl or H-Ala-OAllyl·HCl; Note 11). Essentially the same method is followed as in Method A above. H-Phe-O*t*Bu·HCl (49 mg, 10 Eq.) or H-Ala-OAllyl·HCl (33 mg, 10 Eq.) and NaBH$_3$CN (12 mg, 10 Eq.) are combined, dissolved in DMF (0.6 mL), added to the *o,p*-PALdehyde-Ile-PEG-PS resin (100 mg, 0.19 mmol/g), and reacted at 25°C for 18 h to give H-(BAL-Ile-PEG-PS)Phe-O*t*Bu resin or H-(BAL-Ile-PEG-PS)Ala-OAllyl resin. The resins are washed consecutively with DMF (5 x 0.5 min), CH$_2$Cl$_2$ (3 × 0.5 min), DMF (3 × 0.5 min), piperidine-DMF (1:4, 3 ×1 min), DMF (5 × 0.5 min), and CH$_2$Cl$_2$ (3 × 0.5 min); dried in *vacuo*; and used as a starting point for manual chain assembly of peptides by incorporating further protected amino acids. Yields (95% in both cases) are calculated by amino acid analysis.

Acylation of the Sterically Hindered Secondary α-Amino Group Attached to the BAL-Anchor

Method A

This method is for R^1 = CO$_2$*t*Bu and R^2 = CH$_2$-Ph. Fmoc-Leu-OH (67 mg, 10 Eq.) is dissolved in CH$_2$Cl$_2$–DMF (9:1, 0.5 mL; note 12), DIEA (65 µL, 20 Eq.) is added, and this solution is added to the H-(BAL-Ile-PEG-PS)Phe-O*t*Bu resin. After 30-s stirring, solid N-[(dimethylamino)-1H-1,2,3-triazolo-[4,5-b]pyridin-1-yl-methylene]–N-methylmethanaminium hexafluorophosphate N-oxide (HATU, 72 mg, 10 Eq; note 14) is added to initiate coupling. After 2 h, the dipeptide-resin is washed with DMF (5 × 0.5 min) and CH_2Cl_2 (5 × 0.5 min), and a second 2-h coupling by the same procedure is carried out. Fmoc removal, hydrolysis, and amino acid analysis give a yield of 95%.

Method B

This method is for $R^1 = CO_2$Allyl and $R^2 = CH_3$. Trt-Gly-OH (60 mg, 10 Eq.; note 14) is dissolved in CH_2Cl_2–DMF (9:1; 0.6; note 12 mL), DIEA (65 µL, 20 Eq.) is added, and the solution is added to the H-(BAL-Ile-PEG-PS)Ala-OAllyl resin. Coupling initiated by addition of solid PyAOP (note 14) (99 mg, 10 Eq.) is carried out for 2 h. The peptide-resin is then washed with DMF (5 × 0.5 min) and CH_2Cl_2 (5 × 0.5 min), and the coupling procedure (2 h) is repeated. Acylation yield (95%) is calculated by amino acid analysis.

Incorporation of the Third Protected Amino Acid to Circumvent the Diketopiperazine Side Reaction That Occurs during Syntheses of Cyclic Peptides ($R^1 = CO_2$Allyl) (Note 15)

After method B (just above), trityl removal with TFA–H_2O–CH_2Cl_2 (2:1:97, 5 × 1 min) is followed by washing with CH_2Cl_2 (5 × 0.5 min). Next, Fmoc-Glu(OtBu)-OH (81 mg, 10 Eq.) and PyAOP (99 mg, 10 Eq.) are dissolved separately in DMF (0.6 mL total), combined, and added to the resin. In situ neutralization / coupling initiated by the addition of DIEA (65 µL, 20 Eq.) is carried out for 2 h.

Cleavage

Final products are cleaved with TFA–H_2O (19:1) (1 mL / 50 mg of resin) at 25°C for 1 h. The filtrate from the cleavage reaction is collected, combined with TFA washes (1 mL / 50 mg of resin) of the cleaved peptide-resin, and dried. Cleavage yields (>85%) are calculated by amino acid analysis.

NOTES

1. The chemistry described throughout this chapter is equally successful when working with pure para, pure ortho, or *ortho / para* isomer mixtures.

2. Owing to the viscous reaction mixture, it is necessary to use a mechanical stirrer (magnetic stirring is insufficient).

3. Our experience shows that linker preparation and applications are also successful using $Br(CH_2)_nCO_2Et$, where $n \geq 2$.

4. BAL chemistry is compatible with a wide range of functionalized polymeric supports, including PS, PEG-PS, and Synphase crowns.

5. To accurately determine anchoring, coupling, and cleavage yields, resins are extended further with an internal reference amino acid[2] (IRAA; Ile is used), introduced as its Fmoc derivative by standard coupling methods, at a point before introduction of the handle.

6. Commercial PEG-PS has a Nle IRAA *between* the PS and bifunctional PEG, the latter of which sometimes acts as a spacer and other times cross-links two Nle sites. Hence, ratios of Nle-incorporated amino acids of 2.5–4 represent quantitative yields.

7. Other coupling reagents, such as HBTU or HATU, are also effective in place of BOP/HOBt as described.

8. Unless contraindicated for economic reasons, it is recommended to use 10 Eq. each of amine and $NaBH_3CN$ for the on-resin reductive amination step. In some cases, as little as 1–2 Eq. of amine will give efficient incorporation. $NaBH(OAc)_3$ can be used instead of $NaBH_3CN$. As a rule, reactions should be performed at 25°C.

9. When incorporating an optically active amino acid derivative, a separate imine-forming step should be avoided.

10. Peptide resin samples are hydrolyzed in 12 N aqueous HCl–propionic acid (1:1) at 155°C for 3 h.

11. Other counterions besides chloride, such as trifluoroacetate and tosylate, are also appropriate for these solid-phase reductive aminations.

12. For acylation of a resin-bound secondary amine, the choice of solvent is critical. We find that CH_2Cl_2 or CH_2Cl_2–DMF (9:1) give the optimal results.

13. Alternative methods are described in the original paper and reviewed in "Discussion" below.

14. To decide whether to use Ddz or Trt protection, the following considerations apply: In general, Ddz-protected derivatives couple more efficiently that the corresponding Trt derivatives. Thus, Trt-Gly-OH and Trt-Ala-OH couple very well, but more sterically crowded amino acids with Trt protection couple slowly and Ddz is preferred. However, because Ddz removal conditions require a somewhat higher acid concentration, low-level premature cleavage (1–3%) of dipeptide from the resin can occur as a side reaction.

15. To circumvent diketopiperazine side reactions that occur during syntheses directed at cyclic peptides and peptide

esters ($R^1 = CO_2Allyl, CO_2R^4$), Trt- or Ddz-amino acids are used at the second cycle of incorporation, as explained in "Discussion."

DISCUSSION

Solid-phase synthesis of biomolecules, of which peptides are the prime example, is well established. The search for more effective therapeutic agents creates a need for different strategies to synthesize peptides with C-terminal end groups other than the usual carboxylic acid and carboxamide functionalities. Methods described herein are readily generalized to small nitrogen-containing organic molecules.

In our novel Backbone Amide Linker (BAL)[3] approach for SPS of C-terminal modified peptides, the growing peptide is anchored through the backbone nitrogen instead of through a terminal C^α-carboxyl group, thus allowing considerable flexibility in management of the termini. Initial efforts on BAL have adapted the chemistry of the tris(alkoxy)benzylamide system exploited previously with PAL anchors.[4] The BAL anchor is established by reductive amination of the aldehyde precursors of PAL, e.g., 5-(4-formyl-3,5-dimethoxyphenoxy)valeric acid (**5**) (*p*-PALdehyde) or 5-[(4 or 2)-formyl-3,5-dimethoxyphenoxy]-valeric acid (**5** and **5'**) (*o,p*-PALdehyde), with an amino acid residue (or an appropriately modified derivative), and subsequent *N*-acylation by an appropriately protected second amino acid residue. This gives a dipeptidyl unit that is linked to the support through a backbone amide bond. Further chain growth proceed normally with N^α-9-fluorenylmethoxycarbonyl (Fmoc) solid-phase synthesis protocols. Finally, acidolytic cleavage with trifluoroacetic acid releases the peptide from the resin, with concomitant removal of the side-chain protecting groups.

The first part of this chapter describes the preparation of 4-formyl-3,5-dimethoxyphenol (**2**) (pure isomer) by Vilsmeier formylation of 3,5-dimethoxyphenol (**1**). The phenolic function is alkylated with ethyl 5-bromovalerate, and this intermediate is saponified to the corresponding acid, 5-(4-formyl-3,5-dimethoxy-phenoxy)valeric acid (*p*-PALdehyde) (**5**). Alternatively, the 3,5-dimethoxyphenol (**1**) is alkylated first, followed by Vilsmeier formylation, which provides a mixture of ortho and para isomers. Subsequent steps give the ortho/para mixture 5-[(4 or 2)-formyl-3,5-dimethoxyphenoxy] valeric acid (*o,p*-PALdehyde).[4]

The second part of this chapter describes the quantitative coupling of *p*-PALdehyde or *o,p*-PALdehyde to an amino-functionalized solid support poly(ethylene glycol)-polystyrene graft (PEG-PS)[5] via a BOP/HOBt/DIEA (2:3:3) or HATU/DIEA (1:2) mediated coupling. These procedures yield the *p*-PALde-hyde-resin or *o,p*-PALdehyde-resin. Subsequently, attachment of the C-terminal amino acid residue (with its C^{α}-carboxyl group suitably protected as required or alternatively with an appropriate C-terminal modification) through its amino group is carried out via an on-resin reductive amination procedure using conditions similar to those developed by Sasaki and Coy.[6] Either the free amine or any of a variety of salts (hydrochloride, trifluoroacetate, or tosylate) can be used. Our optimized protocols give the desired BAL anchors in nearly quantitative incorporation (i.e., 95%, as judged by IRAA's)[2] with either MeOH or *N,N*-dimethylforma-mide (DMF) as solvents, and using the amine and cyanoborohy-dride, both in considerable excess (10 Eq. each) over resin-bound aldehyde. (Solvents of choice are DMF[6] or MeOH.[3] Given the tendency for dialkylation in solution with DMF as solvent,[3] the relative absence of such an unfavorable side reaction in the solid-phase case is taken as evidence for relative site isolation. The success of on-resin monoreductive amination in DMF is also attributable to the considerable excess of amine, later removed readily by filtration and washing, which can be used in the reaction.) Our optimal protocols, when applied to amino acid derivatives, proceed without racemization and could be success-fully transferred to other immobilized aldehydes on polymeric supports; the keys to this may be to avoid pre-equilibration and to ensure a neutral or slightly acidic reaction milieu.

Finally, we describe acylation of the sterically hindered secondary α-amino group attached to the BAL-anchor. Commonly applied in situ coupling reagents[7] in DMF—for example, BOP, HATU, and *N*-[(1*H*-benzotriazol-1-yl)(dimethylamino)-methylene]-*N*-methylmethanaminium hexafluorophosphate *N*-oxide (HBTU), used in the equimolar presence of bases such as *N*-methylmorpholine (NMM) or *N*,*N*-diisopropylethylamine (DIEA), and/or additives such as 1-hydroxybenzotriazole (HOBt) or 3-hydroxy-3*H*-1,2,3-triazolo[4,5-*b*]pyridine [1-hydroxy-7-azabenzotriazole (HOAt)]—are all inefficient in mediating the acylation. However, high yields for acylation of the secondary amine are obtained by applying the symmetrical anhydrides of Fmoc-amino acids; the optimal solvent is CH_2Cl_2 (plus whatever amount of DMF is needed for solubility reasons, e.g., CH_2Cl_2–DMF (9:1)), and the reaction does not require base. Other reagents giving satisfactory results with CH_2Cl_2–DMF (9:1) as solvent (*always* preferred over neat DMF or similar solvents such as *N*-methyl-2-pyrrolidinone (NMP)) include HATU / DIEA (1:2), 1,1,3,3-tetramethyl-2-fluoroformamidinium hexafluorophosphate (TFFH) / DIEA (1:2), 7-azabenzotriazol-1-yl-*N*-oxytris(pyrrolidino)phosphonium hexafluorophosphate (PyAOP)/ DIEA (1:2), and bromotris(pyrrolidino)phosphonium hexafluorophosphate (PyBroP)/DIEA (1:2). Preformed acid fluorides are also effective, particularly in the presence of DIEA (1.1 Eq.).

With the C-terminal residue introduced as part of the BAL anchor and the penultimate residue incorporated successfully by the optimized acylation conditions just described, further stepwise chain elongation by addition of Fmoc-amino acids generally proceeded normally by any of a variety of peptide synthesis protocols.

Part of our original vision with BAL was to use allyl chemistry to introduce a third dimension of orthogonality and access cyclic peptides. However, we observed that with BAL-anchored glycyl allyl esters, piperidine-promoted removal of Fmoc at the dipeptidyl level was accompanied by almost quantitative diketopiperazine formation. Such a process is favored by the allyl alcohol leaving group, the sterically unhindered Gly residue, and the BAL secondary amide, which allows the required *cis* transition state. It is important to point out that diketopiperazine formation was not observed with *t*Bu ester protection or with modified endgroups at the C-terminus.

Based on earlier precedents,[8] we expected that the level of diketopiperazine formation could be reduced substantially by using an acidolytically removable N^α-amino protecting group so that the amine endgroup of the BAL-anchored dipeptide would remain protonated until the time for coupling. Experimentally, this is accomplished by: (i) incorporation of the penultimate residue as its N^α-trityl (Trt) derivative; (ii) selective detritylation with TFA–H$_2$O–CH$_2$Cl$_2$ (2:1:97), for 5 min without cleavage of the BAL anchor; and (iii) incorporation of the third residue as its N^α-Fmoc derivative under in situ neutralization/coupling conditions mediated by PyAOP/DIEA in DMF *or* (i′) use of the N^α-2-(3,5-dimethoxyphenyl)propyl[2]oxycarbonyl (Ddz) protected derivative; (ii′) removal of Ddz with TFA–H$_2$O–CH$_2$Cl$_2$ (3:1:96), for 6 min; (iii′) same as (iii).

In conclusion, the BAL method is a novel and general strategy for solid-phase synthesis of peptides and peptide derivatives, is compatible with a wide range of functionalized polymeric supports, and is readily generalizable to other nitrogen-containing molecules.[9]

REFERENCES

1. Kaiser, E.; Colescott, R. L.; Bossinger, C. D.; Cook, P. *Anal. Biochem.* **1970**, *34*, 595.

2. Atherton, E.; Clive, D. L.; Sheppard, R. C. *J. Am. Chem. Soc.* **1975**, *97*, 6584; Matsueda, G. R.; Haber, E. *Anal. Biochem.* **1980**, *104*, 215; and Albericio, F.; Barany, G. *Int. J. Pept. Protein Res.* **1993**, *41*, 307.

3. Jensen, K. J.; Alsina, J.; Songster, M. F. et al., *J. Am. Chem. Soc.* **1998**, *120*, 5441.

4. Albericio, F.; Barany, G. *Int. J. Pept. Protein Res.* **1987**, *30*, 206 and Albericio, F.; Kneib-Cordonier, N.; Biancalana, S. et al., *J. Org. Chem.* **1990**, *55*, 3730.

5. Barany, G.; Albericio, F.; Solé, N. A. et al., In Schneider, C. H., Eberle, A. N., eds., *Peptides 1992: Proceedings of the Twenty-Second European Peptide Symposium*, ESCOM Science Publishers: Leiden, The Netherlands, 1993, p. 267; Zalipsky, S.; Chang, J. L.; Albericio, F.; Barany, G. *React. Polym.* **1994**, *22*, 243; and Barany, G.; Albericio, F.; Kates, S. A.; Kempe, M. In: Harris, J. M.; Zalipsky, S., eds., *Chemistry and Biological Application of Polyethylene Glycol, ACS Symposium Series 680*, American Chemical Society Books: Washington, D.C., 1997, p. 239.

6. Sasaki, Y.; Coy, D. H. *Peptides* **1987**, *8*, 119.

7. Albericio, F.; Carpino, L. A. *Methods Enzymol.* **1997**, *289*, 104 and Humphrey, J. M.; Chamberlin, A. R. *Chem. Rev.* **1997**, *97*, 2243.

8. Gairí, M.; Lloyd-Williams, P.; Albericio, F.; Giralt, E. *Tetrahedron Lett.* **1990**, *31*, 7363 and Alsina, J.; Giralt, E.; Albericio, F. *Tetrahedron Lett.* **1996**, *37*, 4195.

9. Boojamra, C. G.; Burow, K. M.; Ellman, J. A. *J. Org. Chem.* **1995**, *60*, 5742, Boojamra, C. G.; Burow, K. M.; Thompson, L. A.; Ellman, J. A. *J. Org. Chem.* **1997**, *62*, 1240; Gray, N. S.; Kwon, S.; Schultz, P. G. *Tetrahedron Lett.* **1997**, *38*, 1161; and Ngu, K.; Patel, D. V. *J. Org. Chem.* **1997**, *62*, 7088.

CHAPTER THIRTEEN

THE ALLYLSILYL LINKER: SYNTHESIS OF CATALYTIC BINDING OF ALKENES AND ALKYNES TO AND CLEAVAGE FROM ALLYLDIMETHYLSILYL POLYSTYRENE

Submitted by MATTHIAS SCHUSTER and SIEGFRIED BLECHERT

Institut für Organische Chemie, Sekr. C3, Technische Universität Berlin, Straße des 17. Juni 135, D-10623 Berlin, Germany

Checked by SHOMIR GHOSH

Leukosite Inc., 215 First Street, Cambridge, MA, USA 02142

REACTION SCHEMES

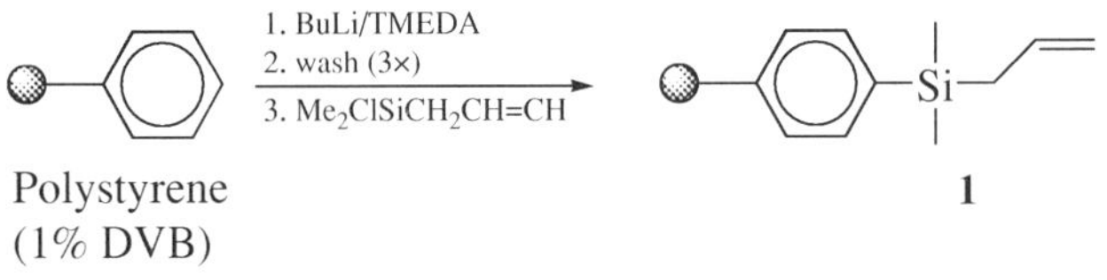
1. BuLi/TMEDA
2. wash (3×)
3. Me2ClSiCH2CH=CH
Si
Polystyrene
(1% DVB)
1

Scheme 1. Synthesis of allyldimethylsilyl polystyrene resin (1% DVB).

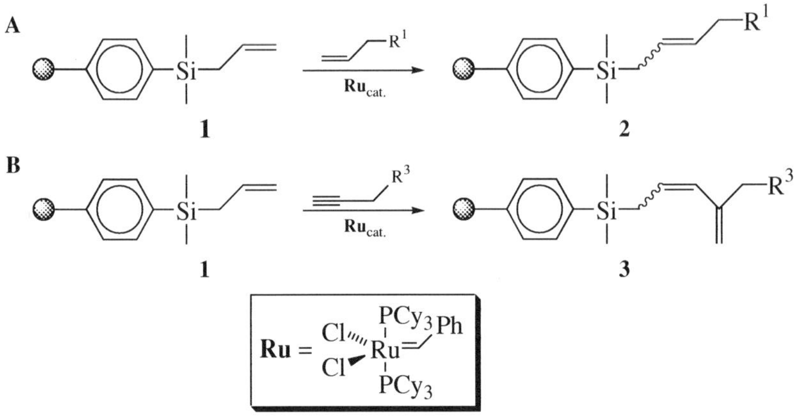

Scheme 2. Catalytic cross-metathesis binding of terminal alkenes (**A**) and alkynes (**B**) to allyldimethylsilyl polystyrene.

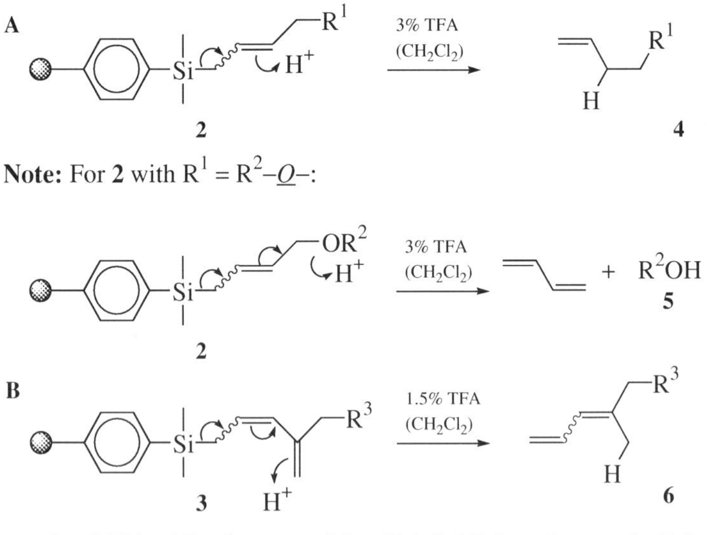

Note: For **2** with $R^1 = R^2-O-$:

Scheme 3. Mild acidic cleavage of the allylsilyl linker via protodesilylation.

PROCEDURES

Preparation of Allyldimethylsilyl Polystyrene Resin (1% DVB)

A two-necked round-bottom flask equipped with a silicone rubber septum and a reverse filter funnel is charged with 8 g polystyrene (1% DVB; note 1) and 120 mL dry cyclohexane; 12 mL (80 mmol) TMEDA and 48 mL (1.6 M in hexane, 77 mmol) BuLi are added, and the suspension is gently shaken for 3 days at ambient temperature under exclusion of moisture and air. The supernatant is removed by reverse filtration under dry nitrogen and replaced by 30 mL dry cyclohexane (note 2). This procedure is repeated twice, and 12 mL (80 mmol) allyldimethylsilyl chloride is added under shaking using a syringe. After 1 h the solvent is removed by reverse filtration under dry nitrogen and 100 mL dimethylformamide is added. After shaking for 10 min, the resin is filtered off; washed repeatedly with methanol, dichloromethane, and diethyl ether; and dried under vacuum.

Catalytic Cross-Metathesis Binding

Terminal Alkenes

Under exclusion of moisture and air (glove box) a 10-mL round bottom flask is charged with 0.3 g **1** (note 3) and 5 mL absolute dichloromethane. Between 0.3 and 0.6 mmol terminal alkene and 12 mg (0.015 mmol) **Ru** (note 4) are added. The resulting suspension is refluxed under stirring for 18 h (glove box). The resin is filtered off and washed with 30 mL *each* of DMF, dichloromethane, methanol, and diethyl ether. Residual diethyl ether is removed under high vacuum.

Terminal Alkynes

Under exclusion of moisture and air (glove box) a 10-mL round bottom flask is charged with 0.3 g **1** and 5 mL absolute dichloromethane; 0.35 mmol terminal alkyne and 12 mg (0.015 mmol) **Ru** are added. The resulting suspension is refluxed under stirring for 18 h (glove box). The resin is filtered off and washed thoroughly as described above and dried under high vacuum.

Cleavage of the Allylsilyl Linker by Protodesilylation

Resins **2** and **3** are treated with dichloromethane containg 3% and 1.5% trifluoroacetic acid (10 mL / g resin), respectively, for 18 h. The resin is filtered off and washed twice with dichloromethane (10 mL / g of resin). The filtrate is washed with saturated $NaHCO_3$ (5 mL) and brine (5 mL), and the organic phase is separated and filtered through a short path silica gel column to obtain a colourless solution. In the case of polymer-bound allyl esters giving rise to cleavage products of type **5f**, the aqueous workup is omitted. The products obtained after removal of solvent under reduced pressure contain small amounts of silanol by-products (note 5), which is to be accounted for in the calculation of cleavage yields.

NOTES

1. Polystyrene (1% DVB) was a kind gift from Bayer AG, Leverkusen. Before use it was repeatedly washed with dichloromethane and diethyl ether and thoroughly dried under vacuum.

2. During the deprotonation, the polystyrene resin takes on a deep red color, which disappears after addition of the silyl chloride.

3. The silicon content of **1** was determined by inductive-coupled plasma-optical emission spectroscopy (ICP-OES) of sodium tetraborate melt samples. It approximated 1 mmol/g resin. Results shown in Tables 13.1 and 13.2 were obtained using a resin containing 1.3 mmol Si per gram of **1**, and results shown in Table 13.3 were obtained using a resin containg 0.9 mmol Si per gram of **1**.

4. Solvents and reagents used were of the highest available purity. **Ru** was obtained from Strem Chemicals, Inc. Allylbenzene and dimethylpropargyl malonate were obtained from

TABLE 13.1. Results of Cleavage of Polystyrene-Supported Allylsilanes 2a–d

Cross-Metathesis Product (**2**)	Cleavage Product (**4**)[a]
$R^1 =$ 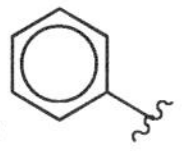 **2a**[b]	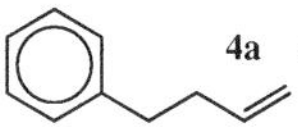 **4a** (0.5 mmol/g)
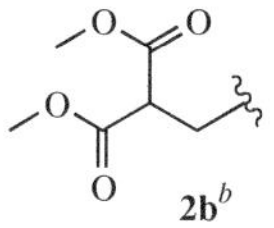 **2b**[b]	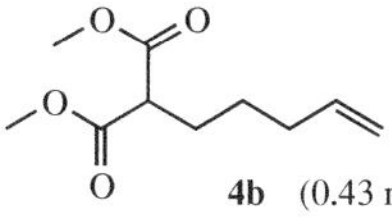 **4b** (0.43 mmol/g)
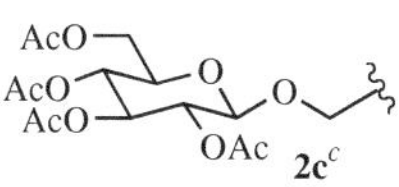 **2c**[c]	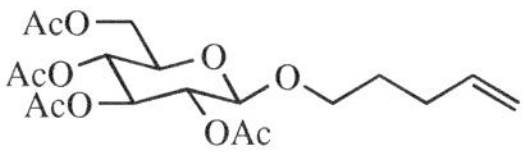 **4c** (0.34 mmol/g)
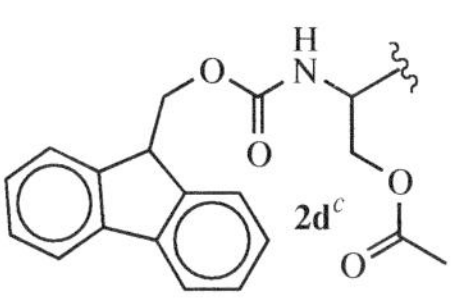 **2d**[c]	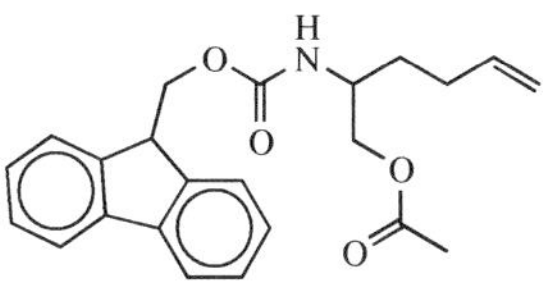 **4d** (< 0.1 mmol/g)

[a] Isolated yield of cleavage product **4** per gram of **2** is given in parentheses.
[b] Metathesis conditions: 300 mg **1**; 0.6 mmol terminal olefin; 0.015 mmol **Ru**; 5 mL CH_2Cl_2 (reflux); 18 h.
[c] Metathesis conditions: 300 mg **1**; 0.3 mmol terminal olefin; 0.015 mmol **Ru**; 5 mL CH_2Cl_2 (reflux); 18 h.

TABLE 13.2. Results of Cleavage of Allylsilanes 2e,f Containing Allyloxy Functions

Cross-Metathesis Product (**2**)	Cleavage Product (**5**)[a]
R^1 = 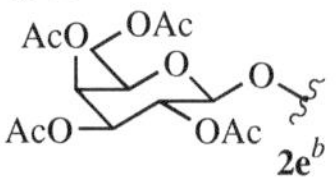 **2e**[b]	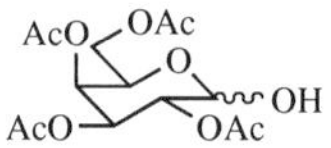 **5e** (0.32 mmol/g)
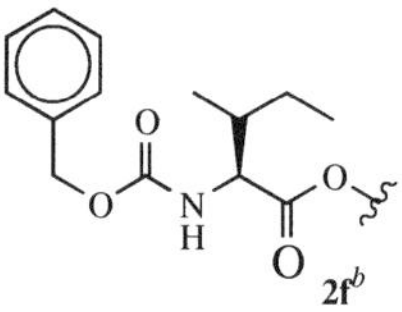 **2f**[b]	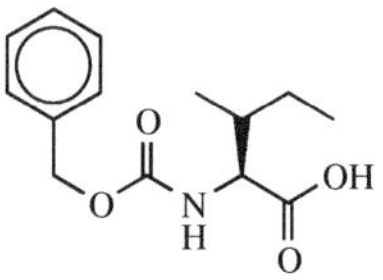 **5f** (0.59 mmol/g)

[a] Isolated yield of cleavage product **5** per gram of **2** is given in parentheses.
[b] Metathesis conditions: 300 mg **1**; 0.3 mmol terminal olefin; 0.015 mmol **Ru**; 5 mL CH_2Cl_2 (reflux); 18 h.

Fluka, and propargyl acetate and propargyl methacrylate were from Lancaster. All other terminal olefins were synthesized according to established standard procedures.

5. During prolonged cleavage homoallyldimethyl silanol is formed as a by-product:

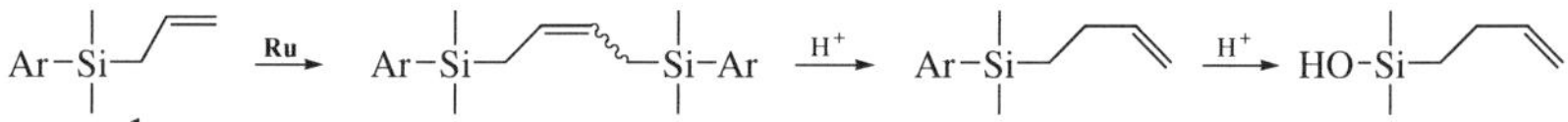

1

It is usually not removed by filtration on silica gel.

DISCUSSION

Olefin metathesis enables the catalytic formation of C=C double bonds under mild conditions.[1] After the development of well-defined catalysts,[1,2] selective cross-couplings between functionalized terminal alkenes (CM) have been noted.[2] A general problem

TABLE 13.3.   Results of Cleavage of Polystyrene-Supported Allylsilanes 3a–f

Cross-Metathesis Product (3)	Cleavage Product (6)[a]
$R^3 =$	
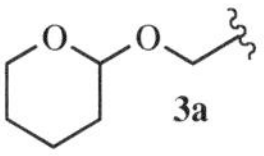 **3a**	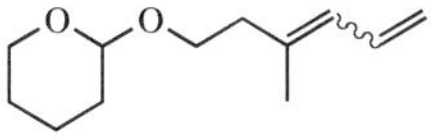 **6a** (0.51 mmol/g, 1:1)
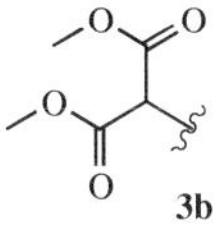 **3b**	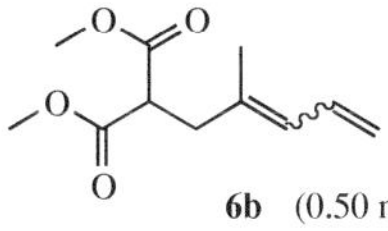 **6b** (0.50 mmol/g, 2:1)
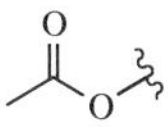 **3c**	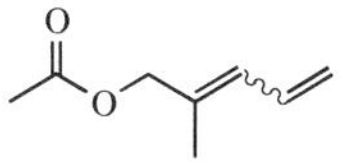 **6c** (0.50 mmol/g, 6:1)
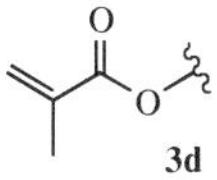 **3d**	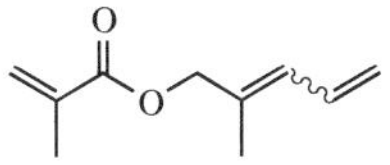 **6d** (0.52 mmol/g, 8:1)
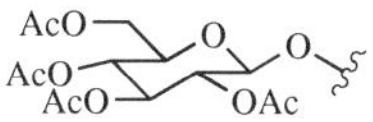 **3e**	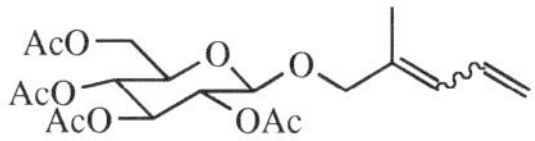 **6e** (0.55 mmol/g, 4:1)

[a] Isolated yield of cleavage product **6** per gram of resin **3** and E/Z-isomer ratio are given in parentheses.

of the crossed metathesis of two different terminal alkenes is the homodimerzation leading to symmetrical cross-products. However, it has been demonstrated, that crossed metatheses of functionalized terminal alkenes with allyltrimethylsilane often proceed in a highly selective manner.[3] The C-Si bond of the resulting functionalized allylsilanes can be cleaved by protodesilylation or fluoride, respectively.[4] When the allylsilane is tethered to the solid support, functionalized olefins can be immobilized by catalytic cross-metathesis. When necessary, the allyl silyl linker can be cleaved under mild acidic conditions. Various terminal olefins have been immobilized using Grubbs's[4] ruthenium carbene initiator **Ru**.[5] The amount of coupled alkene strongly depends on steric parameters. Olefins containing sterically hindered double bonds are not bound. Only alkenes containing functionalities known to be accepted by the catalyst (e.g. esters, acetals, ethers, amides, urethanes) were investigated.

Cleavage of the polymer-bound material was affected by treatment with 3% trifluoroacetic acid in dichloromethane. Two types of products are formed, depending on the structure of resin **2**. Products **4**, containing an additional methylene group compared to the starting alkene, are formed from resins **2** with a carbon atom in the homoallyl position (Table 13.1); whereas protodesilylation of resins **2,** containing an allyl ester or allyl glycoside function, proceeds via a modified mechanism, leading to free carbonic acids or glycosides, respectively (**5** in scheme 3, Table 13.2). The formation of homoallyldimethylsilanol as a by-product of the cleavage reactions (note 5) indicates, that the allylsilyl moieties of **1** partially dimerize on the resin surface during the metathesis reaction.

Only recently a selective crossed metathesis between terminal alkenes and terminal alkynes has been described using the same catalyst.[6] Allyltrimethylsilane proved to be a suitable alkene component for this reaction. Therefore, the concept of immobilizing terminal olefins onto polymer-supported allylsilane was extended to the binding of terminal alkynes. A series of structurally diverse terminal alkynes was reacted with **1** in the presence of catalytic amounts of **Ru**.[7] The resulting polymer-bound dienes **3** are subject to protodesilylation (1.5% TFA) via a conjugate mechanism resulting in the formation of products of type **6** (Table 13.3). Mixtures of *E*- and *Z*-isomers ($E/Z = 8{:}1 - 1{:}1$) are formed. The identity of the dominating *E*-isomer was established by NOE analysis.

In summary, it has been demonstrated, that structurally diverse functionalized alkenes and alkynes are subject to catalytic immobilization onto allylsilyl polystyrene under C,C-bond formation. The allylsilyl linker is cleaved under exceptionally mild acidic conditions.

CHECKER'S COMMENTS

The procedure is reproducible. The yields were lower than that reported. This maybe due to different grades of reagents and solvents used by the checker, further more, a glove box was not used by the checker.

REFERENCES

1. Grubbs, R. H.; Chang, S. *Tetrahedron* **1998**, *54*, 4413; Schuster, M.; Blechert, S.; *Angew. Chem.* **1997**, *109*, 2124 and *Angew. Chem. Int. Ed. Engl.* **1997**, *36*, 2036; and Ivin, K. J.; Mol, J. C. *Metathesis and Metathesis Polymerization*, Academic Press: New York, 1997.

2. Schwab, P.; France, M. B.; Ziller, J. W.; Grubbs, R. H. *Angew. Chem.* **1995**, *107*, 2179 and *Angew. Chem. Int. Ed. Engl.* **1995**, *34*, 2039 and Schrock, R. R.; Murdzek, J. S.; Bazan, G. C. et al. M. *J. Am. Chem. Soc.* **1990**, *112*, 3875.

3. Crowe, W. E.; Goldberg, D. R.; Zhang, Z. J. *Tetrahedron Lett.* **1996**, *37*, 2117 and Brümmer, O.; Rückert, A.; Blechert, S. *Chem. Europ. J.* **1997**, 441.

4. Fleming, I.; Dunogues, J.; Smithers, R. *Org. React.*, **1989**, *37*, 57.

5. Schuster, M.; Lucas, N.; Blechert, S. *Chem. Commun.* **1997**, 823.

6. Stragies, R.; Schuster, M.; Blechert, S. *Angew. Chem.* **1997**, *109*, 2628 and *Angew. Chem. Int. Ed. Engl.* **1997**, *36*, 2518.

7. Schuster, M.; Blechert, S. *Tetrahedron Lett.* **1998**, *39*, 2295.

RESIN-BOUND ISOTHIOCYANATES AS INTERMEDIATES FOR THE SOLID-PHASE SYNTHESIS OF SUBSTITUTED THIOPHENES

Submitted by **HENRIK STEPHENSEN** and
FLORENCIO ZARAGOZA

Novo Nordisk A/S, Novo Nordisk Park, DK-2760 Måløv, Denmark

Checked by **KANG LE** and **ROBERT A. GOODNOW Jr.**

Hoffmann-La Roche Inc., 340 Kingsland Street, Nutley,
New Jersey, USA 07110-1199

REACTION SCHEME

BUILDING BLOCKS

PROCEDURE

Caution! 1,2-Dichloroethane and carbon disulfide are toxic and should be handled only in an efficient hood.

{3-Amino-5-[(3-amino-2,2-dimethylpropyl) amino]-4-(methylsulfonyl)-2-thienyl}(4-biphenylyl)methanone Trifluoroacetate

A fritted polypropylene column is charged with Wang resin–bound 1,3-diamino-2,2-dimethylpropane (note 1) (0.60 g, ca. 0.6 mmol), and the resin is swollen for 1 min in 1,2-dichloroethane (7.0 mL; note 2). The solvent is filtered off; 1,2-dichloroethane (5.2 mL), carbon disulfide (0.8 mL), and diisopropylethylamine (0.52 mL) are added. After shaking for 45 min (note 3) a solution of tosyl chloride (1.32 g, 6.91 mmol) in 1,2-dichloroethane (1.5 mL; note 4) is added to the mixture, and shaking is continued for 15 h. The mixture is filtered, and the resin is washed with dichloromethane (5 × 8.0 mL).

To the product of the previous reaction a solution of methylsulfonylacetonitrile (0.72 g, 6.04 mmol) in DMF (6.0 mL;

note 5) is added, followed by the addition of DBU (0.84 mL). The resulting mixture is shaken for 15 h, filtered, and washed with DMF (5 × 8.0 mL).

To the resin-bound thioamide is added a solution of 4-(bromoacetyl)biphenyl (1.65 g, 6.00 mmol) in DMF (6.0 mL), followed by the addition of acetic acid (0.3 mL; note 6), and the mixture is shaken for 15 h. The mixture is filtered, and the resin washed with DMF (5 × 8.0 mL).

To the product of the previous reaction are added DMF (7.0 mL) and DBU (1.6 mL). After shaking for 15 h, the resin is extensively washed with DMF, dichloromethane and methanol (note 7). Cleavage from the support is effected by treatment with 50% trifluoroacetic acid in dichloromethane (6.0 mL) for 1 h. Concentration of the filtrate yields 273 mg (80%) of the title compound as an oil (85% pure by HPLC, 254 nm), which crystallizes upon addition of methanol (2.0 mL). Filtration and drying yields 82 mg (24%) of slightly yellow crystals, 93% pure by HPLC (254 nm; note 8).

LIBRARY PREPARATION

According to the procedure described above, 27 thiophenes were prepared by combining in all possible ways three symmetric diamines (2,2-dimethyl-1,3-propanediamine, 1,3-propanediamine, 1,4-butanediamine) (note 9); three acceptor-substituted acetonitriles (malonodinitrile, methylsulfonylacetonitrile, (4-chlorophenylsulfonyl)acetonitrile); and three bromoketones ((bromoacetyl)benzene, 4-(bromoacetyl)biphenyl, 2-(bromoacetyl)naphthalene). After cleavage from the support, the purity of the crude products was assessed by HPLC (214 nm, 254 nm) and evaporative light scattering (ELS), and the molecular weight was verified by LCMS. The yield was determined by ^{1}H NMR using DMSO-d_5 as internal standard. The results are listed in Table 14.1.

TABLE 14.1. Results

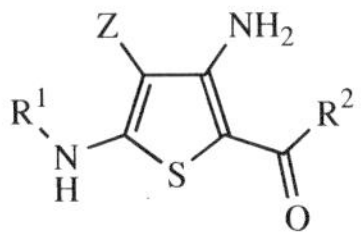

Axx (R^1: 3-amino-2,2-dimethylpropyl)
Bxx (R^1: 3-aminopropyl)
Cxx (R^1: 4-aminobutyl)
xAx (Z: cyano)
xBx (Z: methylsulfonyl)
xCx (Z: (4-chlorophenyl)sulfonyl)
xxA (R^2: phenyl)
xxB (R^2: 4-biphenylyl)
xxC (R^2: 2-naphthyl)

	Purity (RP-HPLC)				
Thiophene	214 nm	254 nm	ELS	Yield	MH^+
AAA	79%	92%	100%	49%	329
AAB	62%	82%	100%	26%	405
AAC	59%	76%	100%	24%	379
ABA	71%	86%	96%	39%	382
ABB	82%	80%	98%	27%	458
ABC	61%	75%	97%	28%	432
ACA	52%	62%	100%	23%	478
ACB	56%	60%	93%	27%	554
ACC	62%	76%	100%	26%	528
BAA	73%	89%	95%	34%	301
BAB	80%	86%	98%	35%	377
BAC	86%	90%	98%	30%	351
BBA	74%	90%	96%	37%	354
BBB	78%	77%	98%	27%	430
BBC	81%	88%	98%	30%	404
BCA	76%	72%	93%	27%	450
BCB	67%	63%	97%	25%	526

TABLE 14.1. (*Continued*)

Thiophene	Purity (RP-HPLC)			Yield	MH^+
	214 nm	254 nm	ELS		
BCC	66%	73%	96%	24%	500
CAA	20%	80%	65%	8%	315
CAB	52%	86%	85%	16%	391
CAC	67%	87%	89%	12%	365
CBA	58%	96%	94%	23%	368
CBB	65%	81%	94%	20%	444
CBC	79%	81%	94%	28%	418
CCA	66%	77%	91%	22%	464
CCB	66%	76%	89%	16%	540
CCC	82%	87%	96%	22%	514

NOTES

1. Prepared from Wang resin (approx. 1 mmol g^{-1}) as described for resin-bound piperazine.[1] We observed that diamines with more than three carbon atoms between the two amino groups lead to unacceptably high degrees of cross-linking ($> 30\%$) when using Wang resin with a loading of 1 mmol g^{-1}. The checkers found that the problem of cross-linking can be minimized by attaching the diamines to 2-chlorotrityl chloride resin (1.34 mmol g^{-1}, Novabiochem).

2. Owing to the mutagenicity of 1,2-dichloroethane, we recommend to replaced this solvent by less hazardous 1,2-dichloropropane. Both solvents are equally suitable for the reactions described herein.

3. Longer reaction times (e.g., 5 h) lead to similar results.

4. If a turbid solution results (precipitation of *p*-toluenesulfonic acid), it might be convenient to filter the solution to avoid plugging of pipettes.

5. Instead of DMF, *N*-methylpyrrolidinone can also be used.

6. Without the addition of acetic acid, the purity of the final product strongly varies. Consistently good results were obtained when the S-alkylation was conducted in the presence of 2–10% acetic acid.

7. Typically, the resin is washed with a mixture of dichloromethane and methanol (2:1; 5 × 10 mL, shaking for 0.5 min each time), with a mixture of dichloromethane (9 mL) and methylamine (1 mL, 30% solution in ethanol), with 1,2-dichloropropane (10 mL) over night, with a mixture of dichloromethane (9 mL) and acetic acid (1 mL; trityl resin–bound products should not be washed with diluted acetic acid), with a mixture of dichloromethane and methanol (2:1; 3 × 10 mL, shaking for 0.5 min each time), and finally with dichloromethane (10 mL). Shorter washing protocols can lead to significant amounts of residual DBU in the final products.

8. Melting point, 216–218°C; IR (KBr) 3459, 3313, 1677, 1549 cm^{-1}; ^{1}H NMR (300 MHz, DMSO-d_6) δ 0.98 (s, 6H), 2.74 (s, 2H), 3.16 (s, 2H), 3.25 (s, 3H), 7.42 (t, J = 7.3 Hz, 1H), 7.52 (t, J = 7.3 Hz, 2H), 7.66–7.80 (m, 6H); ^{13}C NMR (75 MHz, DMSO-d_6) δ 22.25, 35.10, 43.43, 46.02, 54.37, 92.82, 99.29, 126.69, 127.41, 127.94, 128.99, 139.07, 140.08, 141.89, 155.52, 165.93, 183.81. Analysis calculated for $C_{25}H_{28}F_3N_3O_5S_2$ (571.64): C, 52.53; H, 4.94; N, 7.35. Observed: C, 52.60; H, 5.19; N, 7.13.). The checkers found that efficient purification of the crude thiophenes can also be achieved by simple parallel silica gel plug filtration.

9. 2,2-Dimethyl-1,3-propanediamine and 1,3-propanediamine were bound to Wang resin as carbamates (note 1). Because 1,4-butanediamine leads to a high degree of cross-linking

when attached to Wang resin as carbamate, and thereby causing clogging of filters, the trityl resin bound diamine (Novabiochem, 0.40 mmol g^{-1}) was used instead. Each reactor was charged with 100 mg resin (ca. 0.10 mmol).

DISCUSSION

The present procedure[2] describes the conversion of resin-bound, primary aliphatic amines into isothiocyanates and the conversion of the latter into 3-aminothiophenes. The generation of isothiocyanates is related to known procedures,[3] in which amines are first treated with carbon disulfide and the resulting dithiocarbamates are desulfurized by treatment with a condensing agent (alkyl chloroformates, carbodiimides, lead or mercury salts, etc.). The presence of resin-bound isothiocyanates on the polystyrene support could be qualitatively ascertained by infrared spectroscopy (KBr-pellet; strong absorption at 2091 cm^{-1}).

The thiophene synthesis described herein is related to the synthesis in solution reported by Laliberté, and Médawar[4] but differs in some aspects from the procedure in homogeneous phase. Laliberté and Médawar succeeded in obtaining aminothiophenes in a one-pot reaction from acceptor-substituted acetonitriles, isothiocyanates, α-haloketones, and sodium ethoxide. In contrast to their procedure, solid-phase S-alkylation of the intermediate thioamides under basic conditions led to the formation of product mixtures. We obtained pure aminothiophenes only when conducting the S-alkylation under neutral or slightly acidic conditions.

This procedure provides a fast access to substituted thiophenes of sufficient purity to enable direct screening. The synthesis is based on easily available starting materials and can be performed at ambient temperature on standard peptide synthesizers.

REFERENCES

1. Zaragoza, F.; Petersen, S. V. *Tetrahedron* **1996**, *52*, 5999; Dixit, D. M.; Leznoff, C. C. *J. Chem. Soc. Chem. Commun.* **1977** 798; and Dixit, D. M.; Leznoff, C. C. *Israel J. Chem.* **1978**, *17*, 248.

2. Stephensen, H.; Zaragoza, F. *J. Org. Chem.* **1997**, *62*, 6096 and Zaragoza, F. *Tetrahedron Lett.* **1996**, *37*, 6213.

3. Dains, F. B.; Brewster, R. Q.; Olander, C. P. *Org. Synth., Coll. Vol. I*, **1941**, 447; Moore, M. L.; Crossley, F. S. *Org. Synth., Coll. Vol. III*, **1955**, 599; Hodkins, J. E.; Reeves, W. P. *J. Org. Chem.* **1964**, *29*, 3098; and Dowling, L. M.; Stark, G. R. *Biochemistry* **1969**, *8*, 4728.

4. Laliberté, R.; Médawar, G. *Can. J. Chem.* **1970**, *48*, 2709.

AUTHOR INDEX

Albericio, F., **1**, 121
Alsina, J., **1**, 121
Atkinson, G. E., **1**, 85

Barany, G., **1**, 121
Blechert, S., **1**, 139

Chan, W. C., **1**, 85
Cole, D., **1**, 41
Czarnik, A. W., **1**, 15

Dax, S. L., **1**, 9, 45

Ellingboe, J., **1**, 41

Fernandez, M., **1**, 1
Fivush, A. M., **1**, 105
Flygare, J. A., **1**, 1
Fritch, P. C., **1**, 105
Fu, M., **1**, 1

Hamper, B. C., **1**, 55
Hauske, J. R., **1**, 73

Jensen, K. J., **1**, 121

Kearney, P. C., **1**, 1
Kesselring, A., **1**, 55

McNally, J. J., **1**, 9
Mellor, S. L., **1**, 85

Neduvelil, J. G., **1**, 113
Nemeth, G. A., **1**, 101
Nugiel, D. A., **1**, 101

Schuster, M., **1**, 139
Smith, A. L., **1**, 113
Songster, M. F., **1**, 121
Stephensen, H., **1**, 149
Stock, J., **1**, 41

Vagner, J., **1**, 121

Wacker, D. A., **1**, 101
Wang, F., **1**, 73
Willson, T. M., **1**, 105

Xiao, X., **1**, 15

Yang, E., **1**, 15
Youngman, M. A., **1**, 45

Zaragoza, F., **1**, 149
Zhuang, H., **1**, 15

SUBJECT INDEX

Acetals, **1**, 146

Acetic acid, **1**, 10, 47, 67, 129, 152, 155

Acetic anhydride, **1**, 18, 59, 66, 129

Acetone, **1**, 125

Acetonitrile, **1**, 5, 12, 25, 47–48, 60, 69, 82, 87, 152, 156

2-Acetyldiminone, **1**, 97

Acetylenes, **1**, 10, 13

Acidolysis, **1**, 96

Acidolytic cleavage, **1**, 134

Acrylamide resins, **1**, 61–62, 67, 70–71

Acrylate resin, **1**, 55–56, 61, 65, 70–71

Acrylol chloride, **1**, 57–58, 64, 70

Actinonin, **1**, 96

Acylation reactions, **1**, 93, 95, 97, 134, 136

Aldehydes, **1**, 9, 10, 13, 45–46, 49

Allergies, **1**, 6

Alkenes, **1**, 139-140, 145-146

Alkylchloroformates, **1**, 156

Alkynes, **1**, 9, 13, 45-53, 139-140, 142, 146

Allylbenzene, **1**, 143

Allyldimethyl silyl chloride, **1**, 141

Allyldimethyl silyl polystyrene, **1**, 139

Allylsilanes, **1**, 144, 146

Allylsilyl linker, **1**, 139–140

Allyltrimethylsilane, **1**, 145–146

AMEBA resin, **1**, 105–112

Amides, **1**, 111, 121, 146

2-Amidophenols, **1**, 83

Amines, **1**, 2–8, 9–13, 24, 45–46, 49–50, 61, 69–71, 115, 120, 129, 133, 136, 156

Aminobenzoic acid, **1**, 43

2-Amino-4-*tert*-butylphenol, **1**, 80

2-Amino-*p*-cresol, **1**, 80

2-Amino-4-(4-methoxyphenyl)thiazole, **1**, 5

Aminomethyl polystyrene, **1**, 118

2-Aminophenols, **1**, 75-76, 80

2-Aminothiazoles, **1**, 1–8

3-Aminothiophenes, **1**, 156

Angiotensin-converting enzyme (ACE) inhibitors, **1**, 96

Anilines, **1**, 71, 120

Anthranilic acid, **1**, 43, 115–117, 120

Bacterial infections, **1**, 6

Benzaldehyde, **1**, 10, 13

Benzoxazoles, **1**, 73–84

N-Benzyl-β-alanine, **1**, 58

Benzylamine, **1**, 57, 59, 66

p-Benzyloxybenzyl resins, **1**, 41–43, 63

4-(Bromoacetyl)biphenyl, **1**, 152

α-Bromoketones, **1**, 2–3, 6, 152

Bromovalerate, **1**, 125
1,4-Butanediamine, **1**, 155
4-*t*-Butylacetylene, **1**, 13
n-Butyllithium, **1**, 139, 141

Calcium hydride, **1**, 94
Carbamates, **1**, 111, 155–156
Carbodiimide, **1**, 43, 156
Carbon disulfide, **1**, 151, 156
Carbonic acids, **1**, 146
1,1′-Carbonyldiimidazole (CDI), **1**, 80
Cesium, **1**, 43
Chain elongation, stepwise, **1**, 128
4-Chloro-2-amidophenol, **1**, 83
Chloroform, **1**, 63–65
Chloroformate resin, **1**, 117
Chloromethyl aryl solid supports, **1**, 101–104
Chloromethyl polystyrene, **1**, 96
2-Chlorotrityl chloride polystyrene, **1**, 88, 97
2,4,6-Collidine, **1**, 42
Copper (I) chloride, **1**, 9–12, 45, 47, 49
Cyclohexane, **1**, 141
Cyclohexanecarboxaldehyde, **1**, 13
Cyclohexylamine, **1**, 120

trans-1,4-Diaminocyclohexane, **1**, 80
1,3-Diamino-2,2-dimethylpropane, **1**, 151
Diamines, **1**, 75, 152
Dicarboxylic anhydrides, **1**, 75
1,2-Dichloroethane (DCE), **1**, 10, 47, 109–110, 151, 154
Dichloromethane, **1**, 11–12, 17–23, 25, 42, 45, 56-59, 63, 75–76, 79–81, 85, 88–95, 98, 102, 104, 107, 109–110, 114–116, 128–131, 133, 136, 140–142, 146, 151–152, 155
1,2-Dichloropropane, **1**, 154–155
Diethyl azodicarboxylate (DEAD), **1**, 76, 80, 83
Diethyl ether, **1**, 93, 141, 142

N-[(dimethylamino)-1*H*-1,2,3-tria-zolo[4,5-*b*]pyridin-1-ylmethylene]-*N*-methylmethanaminium hexafluorophosphate-*N*-oxide (HATU), **1**, 87, 89–90, 92
Diglycolic acid, **1**, 79
Diglycolic anhydride, **1**, 80
Diisopropylcarbodiimide, **1**, 4, 5
N, N-Diisopropylethylamine (DIEA), **1**, 17–20, 22, 43, 88–90, 92–93, 95, 101, 109–110, 116, 129, 131, 136, 151
p-Dimethoxybenzene, **1**, 6
3,5-Dimethoxyphenol, **1**, 124, 127, 135
N,N-Dimethylacetamide (DMA), **1**, 57–59, 63, 107–108, 110, 116–117
4-(Dimethylamino)pyridine (DMAP), **1**, 76, 79–80
N, N-Dimethylformamide (DMF), **1**, 2–5, 10, 17–20, 24, 42, 46–47, 59, 75–76, 79–80, 85, 89–94, 97, 101–102, 104, 107, 109, 114–117, 124–125, 127–131, 135–136, 141, 151–152
1,4-Dimethylpiperazine, **1**, 9–12, 45, 47, 49
2,2-Dimethyl-1,3-propanediamine, **1**, 155
Dimethylpropargyl malonate, **1**, 143
Dimethlysulfoxide (DMSO), **1**, 6, 11, 24, 57, 59, 61, 66–67, 82, 107, 110
Dioxane, **1**, 3, 5, 9–10, 45, 47
Dithiocarbamates, **1**, 156
n-Dodecylamines, **1**, 67, 69

Endothelin-converting enzyme (ECE) inhibitors, **1**, 96
Enkephalinase inhibitors, **1**, 96
Esters, **1**, 125, 137, 146
1,2-Ethanedithiol, **1**, 93
Ethanol, **1**, 22, 25, 117, 124, 155
Ethers, **1**, 146
Ethyl acetate, **1**, 87–88, 94, 125–128
Ethyl 5-bromovalerate, **1**, 125–127, 135
Ethyl ether, **1**, 17, 19, 22, 124, 127
1-Ethynylcyclohexene, **1**, 13

Fluorenylmethoxycarbonyl (FMOC), **1**, 70, 93-95, 97, 131–132, 134, 136–137
  hydroxylamines, **1**, 87–88, 94, 97
  isothiocyanates (Fmoc-NCS), **1**, 2, 5–6
  chlorides, **1**, 22, 88, 94, 97
  glycines-OH, **1**, 5
Formaldehyde, **1**, 49
4-Formyl-3,5-dimethoxyphenol, **1**, 126, 134
Foroxymithine, **1**, 96

Glycosides, **1**, 146

α-Haloketones, **1**, 156
Hexafluoroisoproponol, **1**, 95
Hexamethyldisiloxane (HMDS), **1**, 63–65, 69
Hexanal, **1**, 13
Hexane, **1**, 88–92, 94, 97, 126
HIV infections, **1**, 6
Homoallyldimethyl silanol, **1**, 144, 146
Homopiperazine, **1**, 80
Hunig's base, see *N, N*-Diisopropyle-thylamine (DIEA)
Hydrazinolysis, **1**, 96
Hydrochloric acid, **1**, 125–126, 133
Hydroxamic acid, **1**, 85–100
1-Hydroxy-7-azabenzotriazole (HOAt), **1**, 87, 89–90, 92
1-Hydroxybenzotriazole (HOBt), **1**, 87, 93, 129, 136
Hydroxylamine hydrochloride, **1**, 87, 97
4-Hydroxy-2-methoxybenzaldehyde, **1**, 107–108, 110–111
Hydroxymethyl-Photolinker AM resin, **1**, 103
Hydroxymethyl-Photolinker NovaSyn TG resin, **1**, 103
N-Hydroxyphthalimide, **1**, 43, 96
Hypertension, **1**, 6

Inflammation, **1**, 6
Isocyanates, **1**, 20, 24–26

Isopropyl alcohol, **1**, 25
Isothiocyanates, **1**, 149, 156

Kaiser test, **1**, 18, 22
Ketones, **1**, 96

Lead salts, **1**, 156
Lithium chloride, **1**, 42–43
Lithium hydride, **1**, 127

Magnesium sulfate, **1**, 88, 125–128
Mannich reactions, **1**, 9–13, 45–53
Matlystatin B, **1**, 96
Matrix metalloprotease inhibitors, **1**, 96
Mercury salts, **1**, 156
Merrifield resins, **1**, 107–108, 110–112
Metalloprotease inhibitors, **1**, 96
Metathesis reaction, **1**, 146
Methanesulfonyl chloride, **1**, 42–43, 101–102
Methanol, **1**, 3–4, 10–11, 17–20, 22, 42, 47, 57–59, 63, 75–80, 89–93, 97, 102, 107, 109, 116, 125–126, 128–129, 135, 141, 152, 155
2–(4-Methoxyphenyl)ethyl amine, **1**, 109–110
4-Methoxyphenylisocyanate, **1**, 23
4-Methoxysulfonyl chloride, **1**, 92
Methylamine, **1**, 155
N-Methyl anthranilic acid, **1**, 117
3-Methyl-2-butanone, **1**, 126
Methylene chloride, **1**, 2–3, 5, 10, 47, 60
3-Methylglutaric anhydride, **1**, 79–80
N-Methylmorpholine (NMM), **1**, 76, 80, 136
N-Methyl-2-pyrrolidinone (NMP), **1**, 136, 155
Methylsulfonyl acetonitrile, **1**, 151
Michael addition, **1**, 61, 67
MicroTubes, **1**, 15–40
Mitsunobu Reaction, **1**, 73–84

Ninhydrin, **1**, 4, 6, 22, 129
Nitrophenylcarbamates, **1**, 24
4-Nitrophenyl-isocyanate, **1**, 23

Oxidation, **1**, 111

Paraformaldehyde, **1**, 13
Pentane, **1**, 128
Peptoids, **1**, 55–72
Phenol, **1**, 6, 22, 43, 125, 127
Phenylisocyanate, **1**, 23
Phosphorous oxychloride, **1**, 124, 127
Piperazine, **1**, 12, 75, 79, 80, 154
Piperazine trityl resin, **1**, 10
Piperidine, **1**, 2–5, 10, 17, 18–20, 47, 89–90, 92–94, 97, 129, 137
Polystyrene, **1**, 21, 42, 115–116, 141–142
Potassium *tert*-butoxide, **1**, 107–108, 110, 125
Potassium carbonate, **1**, 126
Potassium cyanide, **1**, 6, 22
Potassium hydrogen sulfate, **1**, 88
Potassium hydroxide, **1**, 89
1,3-Propanediamine, **1**, 155
Propargyl acetate, **1**, 144
Propargyl amine, **1**, 45–47, 49
Propargyl methacrylate, **1**, 144
Propionic acid, **1**, 83, 133
Propioxatins, **1**, 96
Proteases, **1**, 96
Pyridine, **1**, 22, 76, 79–80, 83

Quinazolinediones, **1**, 113, 118

Reductive alkylation, **1**, 26
Reductive amination, **1**, 111, 123, 128, 133, 135
Ruthenium carbene initiator (Grubb's), **1**, 146
SASRIN resin, **1**, 103, 107, 109–111
Schizophrenia, **1**, 6
Silyl chlorides, **1**, 142
S_N1 reaction, **1**, 97
Sodium acetate, **1**, 127
Sodium borohydride, **1**, 3, 5
Sodium chloride, **1**, 88, 125, 127–128, 142
Sodium cyanoborohydride, **1**, 129–130, 133, 135

Sodium ethoxide, **1**, 156
Sodium hydrogen carbonate, **1**, 87, 142
Sodium hydroxide, **1**, 124–128
Sodium tetraborate, **1**, 143
Sodium triacetoxyborohydride, **1**, 109–110
Succinic anhydride, **1**, 76, 79–80
Sulfonamide, **1**, 108–109, 111

TBTU, **1**, 93
Tentagel resins, **1**, 103
Tetrahydrofuran (THF), **1**, 75–76, 79–80, 83, 94, 97, 104, 107
Thermal cyclization, **1**, 118
Thiazoles, **1**, 3–4
Thioamides, **1**, 156
Thiophenes, **1**, 149–152
Thionyl chloride, **1**, 43
Thiophenes, **1**, 156
Toluene, **1**, 83, 125
p-Toluenesulfonic acid, **1,** 155
p-Toluenesulfonyl chloride, **1**, 109–110, 151
Triethylamine, **1**, 57–59, 64, 107, 110
Trifluoroacetic acid (TFA), **1**, 3–5, 9–11, 21, 23, 25, 45, 47, 60, 63, 64–65, 69–70, 76, 80–82, 87, 90–93, 95, 98, 109–111, 131–132, 134, 140, 142, 146, 152
Triisopropylsilane, **1**, 93
Trimethylorthoformate (TMOF), **1**, 3, 5
Triphenylphosphines, **1**, 80, 83, 103
Triphosgene, **1**, 114–115
Trityl chloride resin, **1**, 46–47, 96

Ureas, **1**, 15–40, 111
Urethanes, **1**, 146

Vilsmeier formylation, **1**, 135

Wang resin, **1**, 41–43, 56–61, 63, 66, 74–76, 80, 101, 103, 151, 154–155
Zinc, **1**, 96